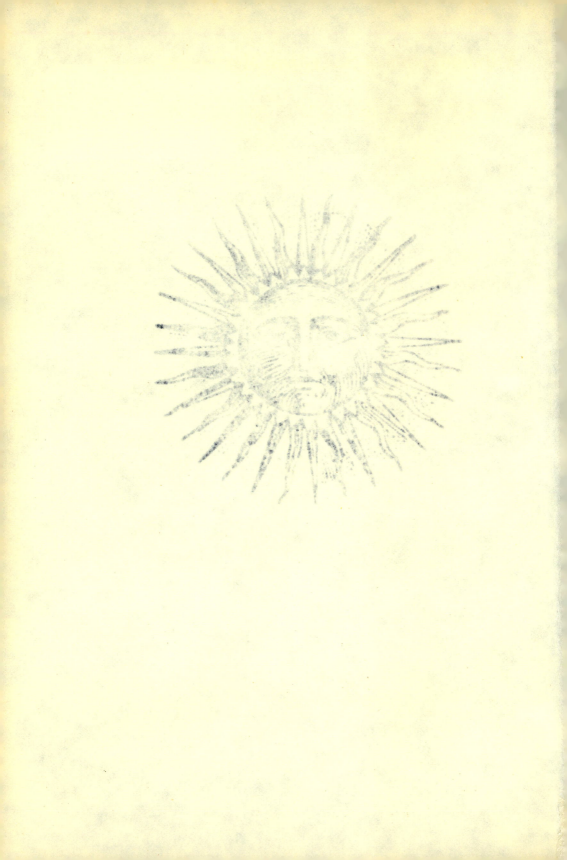

Fundamentals of Statistical Mechanics

Felix Bloch in the classroom, August 1964

FUNDAMENTALS OF STATISTICAL MECHANICS

Manuscript and Notes of
FELIX BLOCH

Prepared by John Dirk Walecka

STANFORD UNIVERSITY PRESS 1989
Stanford, California

Stanford University Press, Stanford, California
© 1989 by the Board of Trustees of the Leland Stanford Junior University
Printed in the United States of America

CIP data appear at the end of the book

Preface

Felix Bloch (1905-1983) was one of the great men of twentieth-century physics. He laid the foundation for the theory of solids and may truly be considered the father of solid-state physics. He played a central role in the theory of magnetism. With Alvarez, he measured the magnetic moment of the neutron. His work on nuclear magnetic resonance was recognized by the award of a Nobel Prize. With Nordsieck, he provided the key insight into the resolution of the infrared problem in quantum electrodynamics. Any one of these accomplishments constitutes a major lifetime achievement. Only a handful of scientists have achieved as many.

Felix Bloch was also one of the giants of Stanford University. He was the first faculty member to receive a Nobel Prize, and for work done at that institution. During his years at Stanford, Felix Bloch taught a course on statistical mechanics many times. In going through his papers, I found problem sets dating back to Spring Quarter of 1933-34. Bloch's first-year graduate course on statistical mechanics was a highlight of the graduate career for several generations of Stanford students.

Statistical mechanics is a powerful subject. It is the key to understanding the physical behavior of the many-body systems making up the world around us. The basic goal is to relate the physical properties of a macroscopic laboratory sample in thermal equilibrium to the detailed dynamical behavior of the atomic and subatomic systems constituting the sample. The number of such constituents is generally vast, comparable to Avogadro's number $N = 6.023 \times 10^{23}/\text{mole}$. Ordinary experience gives us very little feel for this magnitude, yet it provides the basis for equilibrium statistical mechanics. Once the fundamental principles are understood, statistical mechanics immediately relates microscopic behavior to physical observables.

Precisely these fundamental principles lay at the heart of Felix Bloch's course. The principles were initially developed by J. W. Gibbs in his classic work *Elementary Principles of Statistical Mechanics*, published in New York in 1902. Felix was fond of saying that in preparing this course he spent most of his time and effort trying to understand and interpret Gibbs (whom, incidentally, he considered the first truly great American physicist). The course always included applications, but specifically to illustrate the principles. On many occasions, Felix said that it was not a course on applications of statistical mechanics, and a reader who seeks the latest developments in critical behavior, or spin systems, or lattice-gauge theories, is bound to be disappointed.

After retirement, Felix spent the last ten years of his life working on a book based on this course. It was not completed. For one thing, new physics problems continued to absorb his attention. For another, he continued to develop and refine his interpretations and explanations. He was never satisfied, and the complex concepts of coarse-grain and fine-grain averages, of ergodic, quasi-ergodic, and sufficiently ergodic systems, and of the statistical significance of entropy, run through his lectures from the beginning. They were developed and refined each time they were presented. When he discovered a new (and elegant!) algebraic proof of Liouville's Theorem, Felix abandoned the first part of the manuscript and started over, incorporating the new development.

Before his death, Felix Bloch told his son George to use his discretion in the disposition of the existing notes and manuscript. After Felix's death, his wife Lore received inquiries about the possibility of finishing the manuscript and bringing it to publication, and she asked my advice on the matter. I looked over the existing written material and all the old lecture notes that Felix had meticulously written out by hand. The second version of the written text consisted of 72 single-spaced pages, and the coverage of the material extended through thermal averages. There was a long way to go, but the lecture notes were so good, so clear, and so interesting, that I decided I would like to undertake this task myself. In June 1984, Lore and George Bloch, Marvin Chodorow, and I agreed that I should complete the book.

My good friend and colleague Brian Serot had taken the course from Felix in 1976 as a graduate student at Stanford. In Brian's fashion, he had taken a detailed set of notes, and he graciously allowed me to use them in preparing this manuscript. I have done so. My first task was to work through Brian's notes to see the course from the student's point of view. I then worked through all of Felix's lecture notes, the first complete handwritten set dating from 1949, and the most complete and compelling treatments occurring in the 1969 and 1976 versions. I decided to use Bloch's material to construct a one-quarter lecture course. I then wrote up those lectures as a book. In my opinion, three of the detailed developments in the original written manuscript detract from the flow of the lectures, and I have taken the liberty of including this written material verbatim in a series of appendixes. The proof of Liouville's Theorem included in the text is from the 1976 lectures.

Classical statistical mechanics is developed first, and Bloch follows Gibbs' approach utilizing the canonical ensemble. Classical mechanics is reviewed, phase space defined, and Liouville's Theorem proven. The concept of an ensemble of systems is introduced, and thermal equilibrium is related to the constancy of the distribution function. The appropriate averaging in phase space is discussed, and arguments given that lead to Gibbs' canonical ensemble. Energy and work are identified as mean values, and the first and second laws of thermodynamics are used to relate thermodynamic functions to the partition function, which is the desired goal. Applications of classical statistical mechanics include the ideal gas, the virial theorem and non-ideal gases, the equipartition theorem, rotations and vibrations of diatomic molecules, vibrations of solids, electromagnetic radiation in a cavity, and magnetism.

Quantum statistical mechanics is then developed in direct analogy with the classical approach. The basic principles of quantum mechanics are reviewed and the density matrix identified. Quantum phase space is defined in terms of the expansion coefficients in the wave function, and used to introduce the quantum ensemble of systems. This provides a basis for evaluating the statistical average of the density matrix. Thermal equilibrium is related to the constancy of the statistical density matrix. An appropriate average in the quantum phase space again leads to Gibbs' canonical ensemble. A parallel argument then expresses thermodynamic functions in terms of the quantum partition function. The previous examples are revisited in the quantum domain. Finally, the theory of quantum Bose and Fermi gases is developed.

No attempt has been made to include a list of references in the text. (The text references that do appear are meant only to point the reader to a more extensive discussion of a topic at hand.) Rather, a list of selected basic texts and monographs has been included at the end of the book in Appendix E. In that appendix, an attempt has been made to include books or monographs discussing some of the fascinating recent work on statistical mechanics, including phase transitions and critical phenomena, chaos in classical and quantum systems, and lattice gauge theories of quarks and gluons. The interested reader can find further references to pursue in these works.

This book contains over 85 problems, including all the distinct problems assigned by Felix Bloch over the years (a few of which now overlap the text) and some added to supplement the text.

The book assumes a fair degree of sophistication in classical mechanics, although the necessary material is developed in the text. It also assumes a familiarity with the basic concepts of thermodynamics and quantum mechanics. Within this framework, the book should be accessible to anyone in the physical sciences.

Statistical mechanics is unique among topics in physics, involving concepts and principles unlike those that students will meet in their other physics courses. It is important to understand these fundamental principles, which are still under active investigation today, and which form the basis for much of modern research. Statistical mechanics lies at the heart of all areas of natural science, including physics, chemistry, and biology.

If I were to teach statistical mechanics again at either the advanced undergraduate or graduate level, I would use this book and supplement it with one or more books describing selected modern applications. The fundamentals do not change.

My motivation for undertaking this project is that I consider Felix Bloch to be one of the dominant scientific figures of our era, and I feel his insight into the fundamental principles of statistical mechanics is worth sharing with those who are eager to learn. Writing did not come easy to Felix. He agonized over the precise form of everything he put down on paper. It would be presumptuous of me to imagine that I could capture the totality of his thoughts on this subject. I write this book out of respect for the man, and with a firm conviction that others will share my pleasure and satisfaction in his insights. Needless to say, the insights are his, and any misrepresentation or misunderstanding is mine.

This project could not have been completed without the assistance of Lore Bloch, George Bloch, and Marvin Chodorow. Brian Serot's notes were invaluable, as was Sandy Fetter's help. Their contributions and support are gratefully acknowledged.

Finally, I would like to thank Patty Bryant for her dedication and skill in mastering TEX from which this book is printed.

John Dirk Walecka
Professor of Physics
Stanford University
Stanford, California
and
Scientific Director,
CEBAF,
Newport News, Virginia

Contents

Fundamentals of Statistical Mechanics

I INTRODUCTION AND BASIC CONCEPTS(*)

The mechanics of macroscopic systems with many degrees of freedom does not usually allow a rigorous treatment, since the equations of motion are prohibitively difficult to solve in all but a few special cases. Even in these cases, however, one is confronted with the more fundamental difficulty that the information available in practice is far from being sufficient to select the particular solution that best describes a given individual system. In fact, the actual observations normally furnish no more than a small set of macroscopic data, whereas the knowledge of all the many dynamical variables would be required to fully determine the state of the system. With the underlying mechanical process thus largely unspecified, one can account for the observed properties only by resorting to statistical methods.

The foundations of statistical mechanics are due to *Josiah Willard Gibbs* (1839-1903), but their historical roots go back to the development of the kinetic theory of gases. A first major insight was achieved in the eighteenth century by *Daniel Bernoulli* (1700-1782), who realized that the pressure of a gas arises from the impact of the molecules on the walls of the container and thus was able to explain Boyle's law for ideal gases, which states that the pressure is inversely proportional to the volume. The introduction of statistical methods, however, did not occur until about a hundred years later, primarily through the work of *James Clerk Maxwell* (1831-1879) and *Ludwig Eduard Boltzmann* (1844-1906). Maxwell deduced the velocity distribution of molecules of a gas in equilibrium by showing that the distribution remains stationary under the influence of molecular collisions. Boltzmann went further to study the time-dependence of the distribution function and its approach to equilibrium; he also recognized the statistical significance of the concept of entropy in the sense that it is measured by the logarithm of the probability of finding a gas in a given equilibrium state.

A more general problem, however, was raised through the somewhat earlier development of thermodynamics. This subject had been brought to its completed form by *William Thomson* (later Lord Kelvin) (1824-1907) and *Rudolf Emmanuel Clausius* (1822-1888) with the general formulation

---

(*) Sections marked with an asterisk are taken verbatim (with only minor editing) from the handwritten Bloch manuscript.

of the first and second laws and the clarification of such concepts as work, heat, temperature, and entropy. Since the laws of thermodynamics are valid for any macroscopic system, their explanation in mechanical terms required a generalization of the statistical method to encompass all possible cases. Likewise, it is necessary to go beyond the motion of mass points considered in the kinetic theory of gases, and to use as a basis the laws of mechanics in their most general form. The methods of Gibbs are based on the formulation developed by *William Rowan Hamilton* (1805-1865), where a system is characterized by its energy, given as a function of the general coordinates and their conjugate momenta, which determine the equations of motion. Since the work of Gibbs preceded the advent of quantum theory, it refers entirely to classical mechanics, and important modifications are required in the extension to quantum mechanics. The essential statistical elements, however, are common to both; following the historical line, they will first be introduced from the classical viewpoint and later adapted to quantum statistics. In fact, it was the attempt to apply classical statistics to black-body radiation that led *Max Planck* (1858-1947) to the introduction of the quantum of action and, hence, to the beginnings of quantum theory.

II CLASSICAL PHYSICS

1. Hamilton's Equations(*)

The laws of classical mechanics can be formulated by describing a system at a given instant through a set of general coordinates q_i and their conjugate momenta p_i. For a system with "n degrees of freedom," the index $i = 1, 2, \ldots, n$. For N masspoints, one has $n = 3N$; for macroscopic systems n is always a very large number, and in the case of radiation, it is even infinite. The dynamical properties of the mechanical system are specified by the energy expressed as a function of the variables p_i and q_i. This function is called the hamiltonian and shall be denoted by

$$H = H(p_1, \; q_1, \; p_2, \; q_2, \; \ldots, \; p_n, \; q_n) \tag{1.1}$$

or, in abbreviated form,

$$H = H(p_i, \; q_i) \tag{1.2}$$

where the symbols p_i and q_i will be used to indicate the two complete sets of n variables p_i and q_i, respectively. A knowledge of the function $H(p_i, q_i)$ is sufficient to yield the equations of motion in the form of Hamilton's equations

$$\frac{dp_i}{dt} = -\frac{\partial H}{\partial q_i} \qquad ; \; i = 1, \; 2, \; \ldots, \; n$$

$$\frac{dq_i}{dt} = \frac{\partial H}{\partial p_i} \tag{1.3}$$

for the time derivatives of each of the momenta and coordinates. Because one thus deals with a system of first-order differential equations, the values of the dynamical variables at a given time t_0 lead upon integration to their values at any other time t.

The simplest application of these equations is the motion of a non-relativistic mass point in three dimensions. As general coordinates and momenta we take the usual cartesian values

$$
\begin{array}{ll}
q_1 = x & p_1 = p_x \\
q_2 = y & p_2 = p_y \\
q_3 = z & p_3 = p_z
\end{array}
\tag{1.4}
$$

The hamiltonian has the form

$$H = T + V$$

$$= \frac{1}{2m}\mathbf{p}^2 + V(\mathbf{r})$$

$$H = \sum_{i=1}^{3} \frac{p_i^2}{2m} + V(q_1, q_2, q_3)$$

(1.5)

Hamilton's equations for the time derivative of the coordinates then give

$$\frac{dq_i}{dt} = \frac{\partial H}{\partial p_i} = \frac{1}{m} p_i \qquad ; \; i = 1, \; 2, \; 3 \tag{1.6}$$

or

$$p_i = m \frac{dq_i}{dt} \tag{1.7}$$

The equations for the time derivative of the momenta are

$$\frac{dp_i}{dt} = -\frac{\partial V}{\partial q_i} \equiv F_i \qquad ; \; i = 1, \; 2, \; 3 \tag{1.8}$$

which defines the ith component of the force. In vector notation these equations read

$$\mathbf{p} = m\mathbf{v}$$

$$\frac{d\mathbf{p}}{dt} = -\nabla V = \mathbf{F}$$

(1.9)

which are indeed Newton's laws.

As another important special case, Hamilton's equations include those of newtonian mechanics for a material system of N masspoints. With the notation

$$q_{s1} = x_s \qquad ; \; s = 1, \; 2, \; \ldots, \; N$$

$$q_{s2} = y_s \qquad \qquad \text{particle label} \tag{1.10}$$

$$q_{s3} = z_s$$

for the cartesian coordinates and

$$p_{s1} = p_{sx}$$
$$p_{s2} = p_{sy} \tag{1.11}$$
$$p_{s3} = p_{sz}$$

for the corresponding components of the momenta, the pair

$$p_{s\alpha},\ q_{s\alpha} \qquad ;\ s = 1,\ 2,\ \dots,\ N$$

particle label

$$\alpha = 1,\ 2,\ 3$$

(1.12)

cartesian coordinates

thus stands for the pair p_i, q_i ($i = 1, 2, \dots, 3N$) in the previous notation, applied to a system of $n = 3N$ degrees of freedom. The hamiltonian is here the sum

$$H = T + V \tag{1.13}$$

of the kinetic energy $T(p)$ as a function of the momenta $p_{s\alpha}$ and of the potential energy $V(q)$ as a function of the coordinates $q_{s\alpha}$. The square of the momentum for the sth particle is

$$\mathbf{p}_s^2 = \sum_{\alpha=1}^{3} \mathrm{p}_{s\alpha}^2 \tag{1.14}$$

and the kinetic energy is given by the expression

$$T = \sum_{s=1}^{N} \frac{\mathbf{p}_s^2}{2\mathrm{m}_s} \tag{1.15}$$

The potential energy V contains both the interactions between the particles and the effects of the walls. We assume that V has the form

$$V = \sum_{s=1}^{N} V_W(\mathbf{r}_s) + \frac{1}{2} \sum_{s \neq s'=1}^{N} V(|\ \mathbf{r}_s - \mathbf{r}_{s'}\ |) \tag{1.16}$$

where a typical wall potential V_W can be sketched as in Figure 1.1.

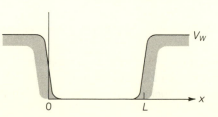

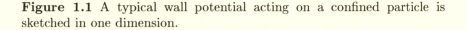

Figure 1.1 A typical wall potential acting on a confined particle is sketched in one dimension.

The components $f_{s\alpha}$ of the force acting on the mass point s are given by

$$f_{s\alpha} = -\frac{\partial V}{\partial q_{s\alpha}} \tag{1.17}$$

Hamilton's equations for the time derivative of the components of momenta in the present notation take the form

$$\frac{dp_{s\alpha}}{dt} = f_{s\alpha} \tag{1.18}$$

They are equivalent to the statement for vectors in Newton's mechanics that the rate of change of momentum is equal to the force. On the other hand, the form

$$\frac{dq_{s\alpha}}{dt} = \frac{\partial T}{\partial p_{s\alpha}} \tag{1.19}$$

of the second set of Hamilton's equations represents the relation between the components of the velocity

$$v_{s\alpha} = \frac{dq_{s\alpha}}{dt} \tag{1.20}$$

and the components $p_{s\alpha}$ of the momentum. With T given by Eq.(1.15), this relation can be written in the form

$$p_{s\alpha} = m_s v_{s\alpha} \tag{1.21}$$

in accordance with Newton's definition of the momentum as the product of mass and velocity. A combination of these results yields

$$m_s \frac{d^2 q_{s\alpha}}{dt^2} = f_{s\alpha} \tag{1.22}$$

as the statement of Newton's law that the force is equal to the mass times the acceleration.

As another application of Hamilton's equations, the previous examples can be extended to describe relativistic motion of mass points if the single-particle kinetic energy appearing in the hamiltonian is taken to have the form

$$T = c(\mathbf{p}^2 + m_0^2 c^2)^{1/2} - m_0 c^2$$

$$= c\left(\sum_{i=1}^{3} p_i^2 + m_0^2 c^2\right)^{1/2} - m_0 c^2 \tag{1.23}$$

where m_0 is the rest mass of the particle and c is the velocity of light. In this case the time derivative of the coordinate is given by

$$\frac{dq_i}{dt} = \frac{\partial T}{\partial p_i} = cp_i \left(\sum_{i=1}^{3} p_i^2 + m_0^2 c^2 \right)^{-1/2} \qquad ; \; i = 1, \, 2, \, 3 \qquad (1.24)$$

or, in vector notation,

$$\mathbf{v} = \frac{c\mathbf{p}}{\sqrt{\mathbf{p}^2 + m_0^2 c^2}} \qquad (1.25)$$

This relation can be squared and inverted to give

$$\mathbf{p}^2 = \frac{m_0^2 \mathbf{v}^2}{1 - \mathbf{v}^2/c^2} \qquad (1.26)$$

The square root of this relation then gives

$$p = \frac{m_0 v}{\sqrt{1 - \mathbf{v}^2/c^2}} \qquad (1.27)$$

The validity of Eq.(1.21) in the relativistic case is seen to demand a dependence of the mass on the velocity such that

$$m \equiv \frac{m_0}{\sqrt{1 - v^2/c^2}} \qquad (1.28)$$

where

$$v^2 \equiv \mathbf{v}^2 = \sum_{i=1}^{3} v_i^2 \qquad (1.29)$$

The equation of motion is now again given by Eqs.(1.8) or (1.18); in the case of many particles, the mass and velocity carry a label s. One can go over to the limiting case of zero rest mass $m_0 = 0$ with the result of the contribution

$$T = cp \qquad (1.30)$$

to the kinetic energy from Eq.(1.23) and velocity

$$v = c \qquad (1.31)$$

from Eq.(1.25).

One advantage of Hamilton's equations is that they are invariant against a large class of transformations. In fact, they follow from a variational principle

$$\delta \int_{t_1}^{t_2} L \, dt = 0 \qquad (1.32)$$

with

$$L \equiv \sum_i p_i \frac{dq_i}{dt} - H(p_i, \ q_i)$$

(1.33)

and fixed endpoints so that

$$\delta p_i(t) = 0 \quad ; \text{ for } t = t_1 \text{ and } t = t_2$$

$$\delta q_i(t) = 0$$

(1.34)

We treat δp_i and δq_i as independent variations. It follows from the Euler-Lagrange equations that

$$-\frac{dq_i}{dt} + \frac{\partial H}{\partial p_i} = 0 \quad ; \ i = 1, \ 2, \ \ldots, \ n$$

$$\frac{dp_i}{dt} + \frac{\partial H}{\partial q_i} = 0$$

(1.35)

Consider a general transformation

$$p'_k = p'_k(p_i, \ q_i) \quad ; \ k = 1, \ 2, \ \ldots, \ n$$

$$q'_k = q'_k(p_i, \ q_i)$$

(1.36)

and an arbitrary function $F(p_i, \ q_i; \ p'_k, \ q'_k)$. Then one also has

$$\delta \int_{t_1}^{t_2} L' dt = 0$$

(1.37)

with L' defined by

$$L' \equiv L + \frac{dF}{dt}$$

(1.38)

Indeed, the variation of F is given by

$$\delta F = \sum_i \left[\frac{\partial F}{\partial q_i} \delta q_i + \frac{\partial F}{\partial p_i} \delta p_i \right] + \sum_k \left[\frac{\partial F}{\partial q'_k} \delta q'_k + \frac{\partial F}{\partial p'_k} \delta p'_k \right]$$

(1.39)

and therefore

$$\delta \int_{t_1}^{t_2} \frac{dF}{dt} dt = \delta F \ |_{t_1}^{t_2} = 0$$

(1.40)

since both δq_i, and δp_i, and also

$$\delta q'_k = \sum_i \left(\frac{\partial q'_k}{\partial p_i} \delta p_i + \frac{\partial q'_k}{\partial q_i} \delta q_i \right) \quad ; \ k = 1, \ 2, \ \ldots, \ n$$

$$\delta p'_k = \sum_i \left(\frac{\partial p'_k}{\partial p_i} \delta p_i + \frac{\partial p'_k}{\partial q_i} \delta q_i \right)$$

(1.41)

vanish for $t = t_1$ and $t = t_2$ [Eqs.(1.34)].

In order that Hamilton's equations hold in terms of the new variables and new hamiltonian

$$\frac{dp'_k}{dt} = -\frac{\partial H'}{\partial q'_k} \qquad ; \; k = 1, \, 2, \, \ldots, \, n$$

$$\frac{dq'_k}{dt} = \frac{\partial H'}{\partial p'_k} \tag{1.42}$$

with

$$H = H[p_i(p'_k, \, q'_k), \, q_i(p'_k, \, q'_k)] \equiv H'(p'_k, \, q'_k) \tag{1.43}$$

it is necessary that

$$L' = \sum_i p_i \frac{dq_i}{dt} - H + \frac{dF}{dt} = \sum_k p'_k \frac{dq'_k}{dt} - H' \tag{1.44}$$

Therefore

$$\frac{dF}{dt} = \sum_k p'_k \frac{dq'_k}{dt} - \sum_i p_i \frac{dq_i}{dt} \tag{1.45}$$

or

$$dF = \sum_k p'_k dq'_k - \sum_i p_i dq_i \tag{1.46}$$

Transformations that leave the form of Hamilton's equations invariant are said to be *canonical*.†

We illustrate these results with a *point transformation*, or coordinate transformation

$$q'_k = q'_k(q_i) \qquad ; \; k = 1, \, 2, \, \ldots, \, n \tag{1.47}$$

which forms a special case of the above. We may look for a transformation with $F = 0$, in which case Eq.(1.46) reduces to

$$\sum_k p'_k dq'_k = \sum_i p_i dq_i = \sum_k \sum_i p_i \left(\frac{\partial q_i}{\partial q'_k} \right) dq'_k \tag{1.48}$$

---

† The function F that *generates* the canonical transformation through Eqs.(1.38) and (1.44) can be generalized to include an explicit dependence on the time t [see A. L. Fetter and J. D. Walecka, *Theoretical Mechanics of Particles and Continua*, McGraw-Hill Book Co., New York (1980)]; this extension is unnecessary for the present development.

This equation is satisfied if the new momenta are defined by

$$p'_k \equiv \sum_i p_i \frac{\partial q_i}{\partial q'_k} \qquad ; k = 1, 2, \ldots, n \qquad (1.49)$$

We give two examples:

1. Transformation to polar coordinates in two dimensions.

Introduce the usual radial and polar coordinates in two dimensions. The cartesian coordinates are given by

$$x = r \, \cos \phi$$
$$y = r \, \sin \phi \qquad (1.50)$$

The new momenta are obtained immediately from Eq.(1.49)

$$p_r = p_x \frac{\partial x}{\partial r} + p_y \frac{\partial y}{\partial r} = p_x \, \cos \phi + p_y \, \sin \phi$$

$$p_\phi = p_x \frac{\partial x}{\partial \phi} + p_y \frac{\partial y}{\partial \phi} = r(-p_x \, \sin \phi + p_y \, \cos \phi) \qquad (1.51)$$

The hamiltonian for a non-relativistic mass point moving in two dimensions then takes the form (in an obvious notation)

$$H = \frac{1}{2m}\left(p_x^2 + p_y^2\right) + V(x, \, y) = \frac{1}{2m}\left(p_r^2 + \frac{p_\phi^2}{r^2}\right) + V(r, \, \phi) \qquad (1.52)$$

Since the transformation is canonical, Hamilton's equations hold in terms of the new general coordinates $(r, \, \phi)$.

2. Transformation to spherical coordinates in three dimensions.

The transformation of the cartesian coordinates to polar-spherical coordinates in three dimensions is given by

$$x = r \, \sin \theta \, \cos \phi$$
$$y = r \, \sin \theta \, \sin \phi \qquad (1.53)$$
$$z = r \, \cos \theta$$

The new momenta are obtained from Eq.(1.49) according to

$$p_r = p_x \frac{\partial x}{\partial r} + p_y \frac{\partial y}{\partial r} + p_z \frac{\partial z}{\partial r} = p_x \sin \theta \cos \phi + p_y \sin \theta \sin \phi + p_z \cos \theta$$

$$p_\phi = p_x \frac{\partial x}{\partial \phi} + p_y \frac{\partial y}{\partial \phi} + p_z \frac{\partial z}{\partial \phi} = r(-p_x \sin \theta \sin \phi + p_y \sin \theta \cos \phi) \qquad (1.54)$$

$$p_\theta = p_x \frac{\partial x}{\partial \theta} + p_y \frac{\partial y}{\partial \theta} + p_z \frac{\partial z}{\partial \theta} = r(p_x \cos \theta \cos \phi + p_y \cos \theta \sin \phi - p_z \sin \theta)$$

A combination of these results gives

$$p_x^2 + p_y^2 + p_z^2 = p_r^2 + \frac{p_\phi^2}{r^2 \sin^2 \theta} + \frac{p_\theta^2}{r^2} \tag{1.55}$$

Thus the hamiltonian for a non-relativistic mass point moving in three dimensions takes the form

$$H = \frac{1}{2m} \left(p_r^2 + \frac{p_\theta^2}{r^2} + \frac{p_\phi^2}{r^2 \sin^2 \theta} \right) + V(r, \theta, \phi) \tag{1.56}$$

Again, since the transformation is canonical, Hamilton's equations hold in terms of the new general coordinates (r, θ, ϕ).

Irrespective of the system considered, it is an important general consequence of Hamilton's equations that the energy is a constant of the motion. Indeed, due to the time-dependence of the momenta and coordinates one has

$$\frac{dH(p_i, q_i)}{dt} = \sum_i \left(\frac{\partial H}{\partial p_i} \frac{dp_i}{dt} + \frac{\partial H}{\partial q_i} \frac{dq_i}{dt} \right) \tag{1.57}$$

and hence from Eqs.(1.3)

$$\frac{dH}{dt} = \sum_i \left(-\frac{\partial H}{\partial p_i} \frac{\partial H}{\partial q_i} + \frac{\partial H}{\partial q_i} \frac{\partial H}{\partial p_i} \right)$$

$$\frac{dH}{dt} = 0 \tag{1.58}$$

In the preceding case dealing with a system of masspoints, this property for the hamiltonian in Eqs.(1.13-1.16) arises from the fact that Eq.(1.17) demands that the force be derived from the gradient of the potential energy. On the other hand, Newton's law Eq.(1.22) remains valid, for example, upon inclusion of a frictional force proportional to the velocity, which leads to a dissipation of energy in violation of Eq.(1.58). It might seem, therefore, that Hamilton's equations are not sufficiently general to allow for dissipative processes. One has to remember, however, that any such process is accompanied by the development of *heat*. In fact, the macroscopic conservation of energy, postulated by the first law of thermodynamics, is in essence based upon the recognition of heat as a form of energy. Regarded, instead, from the microscopic point of view, the purely mechanical energy remains conserved even in frictional processes, with the difference, however, that here it goes over from an ordered to a disordered form in the nature of molecular motion. One of the important goals of statistical mechanics is to generally clarify in what sense a part of the total mechanical energy is understood to correspond to the macroscopic concept of heat.

Both in thermodynamics and in pure mechanics the energy of a system remains constant in the absence of external influences, but can otherwise be made to increase or decrease. Whereas the first law postulates such a change of energy to be additively composed of the amount of heat transferred to the system and of the work performed upon it, only the latter point can be readily expressed in mechanical terms. Indeed, Eq.(1.58) loses its validity if the hamiltonian depends not only on the dynamical variables, but also explicitly upon the time so that Eq.(1.2) is to be replaced by

$$H = H(p_i, \ q_i, \ t) \tag{1.59}$$

With the time derivatives of p_i and q_i still given by Eqs.(1.3), it now follows that instead of Eq.(1.58) one has

$$\frac{dH}{dt} = \frac{\partial H}{\partial t} \tag{1.60}$$

Specifically, this result applies when the hamiltonian is considered as a function of the internal variables p_i and q_i as well as of some external parameters, to be denoted by ξ_j $(j = 1, 2, \ldots, m)$. Writing thus

$$H = H(p_i, \ q_i, \ \xi_j) \tag{1.61}$$

where ξ_j stands for the set of parameters ξ_j $(j = 1, 2, \ldots, m)$, and letting these parameters depend upon the time, one obtains from Eq.(1.60)

$$\frac{dH}{dt} = \sum_{j=1}^{m} \frac{\partial H}{\partial \xi_j} \frac{d\xi_j}{dt} \tag{1.62}$$

The differential change of H during the time interval dt can thus be written in the form

$$dH = \delta W \tag{1.63}$$

where

$$\delta W = \sum_{j=1}^{m} \frac{\partial H}{\partial \xi_j} d\xi_j \tag{1.64}$$

is to be recognized as the differential work performed by external agents in order to change the parameters by the amount $d\xi_j$. The symbol δW, rather than dW, is used in Eqs.(1.63-1.64) to indicate that the righthand side of Eq.(1.64) does not represent a total differential of the hamiltonian given by Eq.(1.61). The appearance of work as force times distance can be formally maintained by considering the quantities ξ_j as coordinates external to the system and the quantities

$$f_j \equiv -\frac{\partial H}{\partial \xi_j} \tag{1.65}$$

as the components of a generalized force against which the work has to be performed, so that one can write

$$\delta W = -\sum_{j=1}^{m} f_j d\xi_j \tag{1.66}$$

As an example, one may consider a single parameter ξ to signify the position of a movable piston of area A, used to vary the volume V of a gas in a cylindrical container (Figure 1.2). In the hamiltonian of Eqs.(1.13-1.16), the effect of the piston, like that of the fixed walls of the container, is described by a sum of terms V_W in the potential energy that increase sharply with small distances of a molecule from the inner surface (Figure 1.1). The parameter ξ here locates the position of the potential V_W due to the surface of the piston. The molecules in the close vicinity of this surface are repelled, and the forces of reaction result in a total force f, exerted from the inside upon the piston. With the definition $p \equiv f/A$ of the pressure as force per unit area and with $dV = Ad\xi$ for the differential change of volume, the expression $\delta W = -fd\xi$ from Eq.(1.66) assumes the familiar form

$$\delta W = -pdV \tag{1.67}$$

for the differential work required to change the volume of the gas.

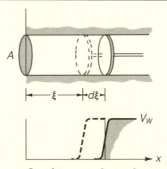

Figure 1.2. A gas is confined to a volume by a movable piston. Work is achieved by the change of an external parameter, in this case the position of the piston ξ, which locates the position of the wall potential V_W.

2. Phase Space(*)

A geometrical interpretation of Hamilton's equations is obtained by considering a set of values of the dynamical variables p_i and q_i ($i = 1, 2, ..., n$) as the coordinates of a point P in a $2n$-dimensional space, called the "phase space."

$$(p_1, q_1, p_2, q_2, \ldots, p_n, q_n) \equiv \text{ point } P$$
$$\text{in the } 2n\text{-dimensional phase space}$$

(2.1)

For a system with a single degree of freedom where $n = 1$, phase space can be visualized and represented by a plane (Figure 2.1); although this is not possible for larger n, this representation serves to symbolize the general case. The notation p, q shall be generally understood to stand for the whole set p_i, q_i ($i = 1, 2, \ldots, n$).

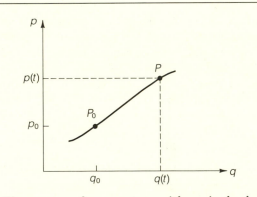

Figure 2.1 Phase space for a system with a single degree of freedom, $n=1$.

Let P_0 represent the point in phase space corresponding to the values p_0, q_0 at the time $t = t_0$. Then, by forward or backward integration of Hamilton's equations, the coordinates p, q of the point P at any other time t are then determined; with t as a continuous variable, the functions $p(t), q(t)$ can be understood to describe a "curve" in phase space, called the *phase orbit*. From the point of view of pure mechanics, the knowledge of these functions or, geometrically, of the motion along the phase orbit, represents, in fact, all that can be said about an individual system. The phase orbit is not necessarily a closed curve, and since a given set of initial values p_0, q_0 completely specifies the subsequent motion of a given mechanical system, phase orbits cannot cross.

It would be necessary, however, to have complete information about the system in the sense that all the values p_0, q_0 are known, or, equivalently, all the values p, q at some other time, in order to determine the motion along a given phase orbit. In practice, this is not normally the case for a macroscopic system where one measures only a few quantities. In thermodynamics, one may not describe a system by more than, say, its pressure and volume or temperature. Suppose the system is composed of $N \cong 10^{23}$ particles. This leaves the very large number $2n = 2 \times 3N$ of data, necessary for the mechanical description, practically unknown and, hence, open to guessing. One is therefore reduced to statistical methods, that is, to a statement of probabilities. In dealing with a mechanical system, it is then necessary, in principle, to consider all possible sets of values p, q that the system may assume at a given time.

Before introducing these statistical methods, however, we prove an important result of pure mechanics, which is known as Liouville's Theorem.

3. Liouville's Theorem(*)

We have seen that the dynamical variables p_i and q_i with $i = 1, 2, \ldots, n$ may be regarded as the coordinates in a $2n$-dimensional space, called the *phase space*. Every set of these variables is thus represented by a point and the quantity

$$d\lambda = \prod_{i=1}^{n} dp_i dq_i \qquad (3.1)$$

represents the *volume element* in the phase space. A mechanical system found at the point $P(p_i, q_i)$ at the time t will follow a definite phase orbit so that at a later time t' it will be found at another point $P'(p'_k, q'_k)$ with the set (p'_k, q'_k) obtained from the initial set (p_i, q_i) by forward integration of Hamilton's equations. Since the new values are determined by the initial values in this fashion, we can write

$$\begin{aligned} p'_k &= p'_k(p_i, q_i) \qquad ; k = 1, 2, \ldots, n \\ q'_k &= q'_k(p_i, q_i) \end{aligned} \qquad (3.2)$$

These equations can be considered as a transformation of the set of variables (p_i, q_i) to the set (p'_k, q'_k) or, geometrically, as a mapping of the phase space at the time t to that at the time t'. The volume element $d\lambda$ and the corresponding volume element $d\lambda'$, obtained through this transformation (Figure 3.1), are related by the expression

$$d\lambda' = J(t, t')d\lambda \qquad (3.3)$$

where $J(t, t')$ is the *jacobian determinant* of the transformation

$$
J(t, t') = \left| \frac{\partial(p'_k, q'_k)}{\partial(p_i, q_i)} \right| \equiv
\begin{vmatrix}
\frac{\partial p'_1}{\partial p_1} & \frac{\partial q'_1}{\partial p_1} & \frac{\partial p'_2}{\partial p_1} & \frac{\partial q'_2}{\partial p_1} & \cdots & \cdots & \frac{\partial p'_n}{\partial p_1} & \frac{\partial q'_n}{\partial p_1} \\
\frac{\partial p'_1}{\partial q_1} & \frac{\partial q'_1}{\partial q_1} & \cdots & & & & & \\
\vdots & & & & & & & \\
\frac{\partial p'_1}{\partial p_n} & \frac{\partial q'_1}{\partial p_n} & \cdots & & & & \frac{\partial p'_n}{\partial p_n} & \frac{\partial q'_n}{\partial p_n} \\
\frac{\partial p'_1}{\partial q_n} & \frac{\partial q'_1}{\partial q_n} & \cdots & & & & \frac{\partial p'_n}{\partial q_n} & \frac{\partial q'_n}{\partial q_n}
\end{vmatrix}
$$

$$(3.4)$$

First, let $t' = t$, so that $p'_k = p_k$, $q'_k = q_k$, and therefore

$$
\frac{\partial p'_k}{\partial p_i} = \delta_{ik} \qquad \frac{\partial q'_k}{\partial q_i} = \delta_{ik}
$$

$$
\frac{\partial p'_k}{\partial q_i} = 0 \qquad \frac{\partial q'_k}{\partial p_i} = 0
$$

$$(3.5)$$

where δ_{ik} is the Kronecker delta defined by

$$
\delta_{ik} = 1 \qquad for\ i = k
$$
$$
= 0 \qquad for\ i \neq k
$$

$$(3.6)$$

In this case all terms on the diagonal of J are equal to unity and all others are zero. Therefore

$$
J(t, t) = 1 \tag{3.7}
$$

Next, let $t' = t + dt$, and keep only first-order terms in dt. One then has from Hamilton's equations

$$
\dot{p}_k \equiv \frac{dp_k}{dt} = -\frac{\partial H}{\partial q_k}
$$

$$
\dot{q}_k \equiv \frac{dq_k}{dt} = \frac{\partial H}{\partial p_k}
$$

$$(3.8)$$

and thus

$$
p'_k = p_k - \frac{\partial H}{\partial q_k} dt
$$

$$
q'_k = q_k + \frac{\partial H}{\partial p_k} dt
$$

$$(3.9)$$

These relations may be differentiated to give

$$
\frac{\partial p'_k}{\partial p_i} = \delta_{ik} - \frac{\partial^2 H}{\partial p_i \partial q_k} dt \quad ; \quad \frac{\partial q'_k}{\partial p_i} = \frac{\partial^2 H}{\partial p_i \partial p_k} dt
$$

$$
\frac{\partial p'_k}{\partial q_i} = -\frac{\partial^2 H}{\partial q_i \partial q_k} dt \quad ; \quad \frac{\partial q'_k}{\partial q_i} = \delta_{ik} + \frac{\partial^2 H}{\partial q_i \partial p_k} dt
$$

$$(3.10)$$

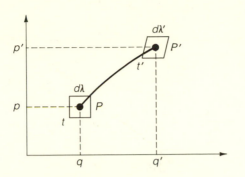

Figure 3.1 The time development of the volume element in phase space as described by Liouville's Theorem (schematic).

In evaluating the determinant in Eq.(3.4), each term containing a non-diagonal element as a factor contains at least another non-diagonal element as a factor. Since all these elements are proportional to dt, these terms are therefore at least of second order in dt and do not contribute to first order. To that order there remains then only the contribution from the diagonal elements

$$\prod_k \left(1 - \frac{\partial^2 H}{\partial p_k \partial q_k} dt\right) \left(1 + \frac{\partial^2 H}{\partial p_k \partial q_k} dt\right)$$

$$\cong 1 + \sum_k \left(\frac{\partial^2 H}{\partial p_k \partial q_k} - \frac{\partial^2 H}{\partial p_k \partial q_k}\right) dt = 1 \qquad (3.11)$$

Therefore

$$J(t, t+dt) = J(t, t) + \frac{\partial}{\partial t} J(t'', t)|_{t''=t} dt = 1 \qquad (3.12)$$

and since Eq.(3.7) states that $J(t, t) = 1$ we have the result

$$\frac{\partial}{\partial t} J(t'', t)|_{t''=t} = 0 \qquad (3.13)$$

On the other hand, besides t and t', consider an arbitrary third time t'' and the corresponding volume element $d\lambda''$ at the time t'' (Figure 3.2) and use Eq.(3.3) with arbitrary t and t'. Then, going from t to t'' we have

$$d\lambda'' = J(t, t'') d\lambda \qquad (3.14)$$

and from t'' to t'

$$d\lambda' = J(t'', t')d\lambda'' \tag{3.15}$$

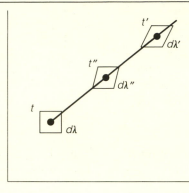

Figure 3.2 Intermediate time t'' and volume element $d\lambda''$ used in proof of Liouville's Theorem.

Therefore

$$d\lambda' = J(t, t'')J(t'', t')d\lambda = J(t, t')d\lambda \tag{3.16}$$

or

$$J(t, t') = J(t, t'')J(t'', t') \tag{3.17}$$

Hence, upon differentiation with respect to the final time t'

$$\frac{\partial J(t, t')}{\partial t'} = J(t, t'')\frac{\partial}{\partial t'}J(t'', t') \tag{3.18}$$

Now on the right side let $t'' = t'$, and this relation becomes

$$\frac{\partial J(t, t')}{\partial t'} = J(t, t')\frac{\partial}{\partial t'}J(t'', t')|_{t''=t'} \tag{3.19}$$

Since Eq.(3.13) holds for any time t, it also holds for t'

$$\frac{\partial}{\partial t'}J(t'', t')|_{t''=t'} = 0 \tag{3.20}$$

Equation (3.19) thus implies

$$\frac{\partial}{\partial t'}J(t, t') = 0 \tag{3.21}$$

This relation now holds for arbitrary t' and provides a first-order differential equation for the jacobian determinant. For $t' = t$ one has the initial condition $J(t,t) = 1$ of Eq.(3.7), and hence

$$J(t,t') = 1 \qquad (3.22)$$

for arbitrary t'. It follows from Eq.(3.3) that

$$d\lambda' = d\lambda \qquad (3.23)$$

One therefore concludes that *the magnitude of the volume element in phase space does not change in the course of its motion along the phase orbits.*

The same result holds for the volume of any closed finite region R in phase space followed along the phase orbits (Figure 3.3). To every volume element $d\lambda$ within that region at the time t, there corresponds an equal volume element $d\lambda'$ within the region R' at the time t' obtained from R by the equations of motion. Integrating the righthand side of Eq.(3.23) over R corresponds to the integration of the lefthand side over R' so that

$$\int_{R'} d\lambda' = \int_R d\lambda \qquad (3.24)$$

or

$$\Omega' = \Omega \qquad (3.25)$$

where Ω and Ω' represent the volume contained within R and R' respectively. In abbreviated form, this result is stated by Liouville's Theorem: The phase volume is a constant of the motion.

This theorem can be illustrated for the simplest case of a mass point in free, one-dimensional motion. The hamiltonian is

$$H = \frac{p^2}{2m} \qquad (3.26)$$

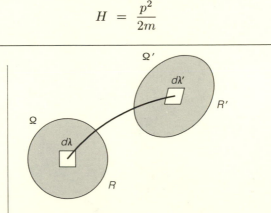

Figure 3.3 Invariance of finite volume elements in phase space followed along the phase orbits (schematic).

and Hamilton's equations give†

$$\dot{p} = -\frac{\partial H}{\partial q} = 0$$

$$\dot{q} = \frac{\partial H}{\partial p} = \frac{p}{m}$$

(3.27)

Integration of these equations yields

$$p = \text{constant}$$

$$q = \frac{p}{m}t + \text{constant}$$

(3.28)

which provides a parametric representation of the phase orbit. In the previous notation we have

$$p' = p$$

$$q' = q + \frac{p}{m}(t' - t)$$

(3.29)

The jacobian of this transformation can be immediatly evaluated as

$$J(t,t') = \begin{vmatrix} \frac{\partial p'}{\partial p} & \frac{\partial q'}{\partial p} \\ \frac{\partial p'}{\partial q} & \frac{\partial q'}{\partial q} \end{vmatrix} = \begin{vmatrix} 1 & \frac{t'-t}{m} \\ 0 & 1 \end{vmatrix} = 1$$

(3.30)

which illustrates our general result. Consider further the rectangle $abcd$ at the time t and the corresponding parallelogram $a'b'c'd'$ at the time t' shown in Figure 3.4. Evidently

$$q_{a'} = q_a + \frac{p_a}{m}(t' - t) \qquad ; \; p_a = p_b$$

$$q_{b'} = q_b + \frac{p_b}{m}(t' - t) \qquad ; \; p_c = p_d$$

$$q_{c'} = q_c + \frac{p_c}{m}(t' - t)$$

$$q_{d'} = q_d + \frac{p_d}{m}(t' - t)$$

(3.31)

Thus

$$q_{b'} - q_{a'} = q_b - q_a \equiv w$$

$$q_{d'} - q_{c'} = q_d - q_c \equiv w$$

(3.32)

---

† We use the customary notation where a dot above a symbol denotes the total time derivative.

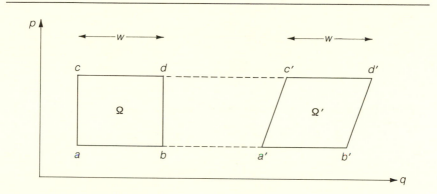

Figure 3.4 Phase volume for free one-dimensional motion of a mass point.

Since $a'b'c'd'$ has the same width w as $abcd$ and the same height $p_c - p_a$, the areas Ω and Ω' are equal, illustrating our general result. The extension to arbitrary regions R and R' (see Figure 3.5) is immediately obtained by dividing the regions into infinitesimally thin horizontal strips and applying the above.

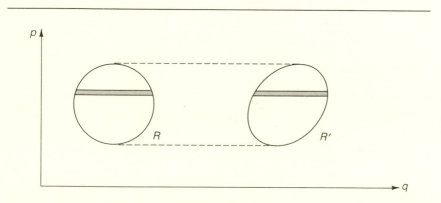

Figure 3.5 Extension of result in Figure 3.4 to a region of arbitrary shape.

Another example is the one-dimensional harmonic oscillator with hamiltonian

$$ H \;=\; \frac{p^2}{2m} \;+\; \frac{1}{2}m\omega^2 q^2 \tag{3.33} $$

Hamilton's equations give

$$\dot{p} = -\frac{\partial H}{\partial q} = -m\omega^2 q$$

$$\dot{q} = \frac{\partial H}{\partial p} = \frac{p}{m} \tag{3.34}$$

A combination of these relations yields

$$\frac{d}{dt}(p + im\omega q) = \dot{p} + im\omega\dot{q} = -m\omega^2 q + i\omega p$$

$$= i\omega(p + im\omega q) \tag{3.35}$$

and hence, by integration

$$p' + im\omega q' = (p + im\omega q)e^{i\omega(t'-t)} \tag{3.36}$$

The real and imaginary parts of this relation are

$$p' = p\,\cos\,\alpha\,-\,m\omega q\,\sin\,\alpha$$

$$q' = \frac{p}{m\omega}\sin\,\alpha\,+\,q\,\cos\,\alpha \tag{3.37}$$

with

$$\alpha \equiv \omega(t' - t) \tag{3.38}$$

the jacobian can be evaluated as

$$J(t,t') = \begin{vmatrix} \frac{\partial p'}{\partial p} & \frac{\partial q'}{\partial p} \\ \frac{\partial p'}{\partial q} & \frac{\partial q'}{\partial q} \end{vmatrix} = \begin{vmatrix} \cos\,\alpha & \frac{\sin\,\alpha}{m\omega} \\ -m\omega\,\sin\,\alpha & \cos\,\alpha \end{vmatrix} = \cos^2\alpha + \sin^2\alpha = 1 \tag{3.39}$$

in accord with our previous analysis. Further, upon change of scale

$$q \rightarrow m\omega q$$

$$q' \rightarrow m\omega q' \tag{3.40}$$

in phase space one has for the area of an arbitrary region in phase space

$$\Omega \rightarrow m\omega\Omega$$

$$\Omega' \rightarrow m\omega\Omega' \tag{3.41}$$

Now it is evident from Eq.(3.37) that the region R' with area $m\omega\Omega'$, which evolves from the region R with area $m\omega\Omega$, is obtained from R through a rotation around the origin by the angle α (Figure 3.6). Since this does not change the area, one has

$$m\omega\Omega' \; = \; m\omega\Omega \tag{3.42}$$

and hence

$$\Omega' \; = \; \Omega \tag{3.43}$$

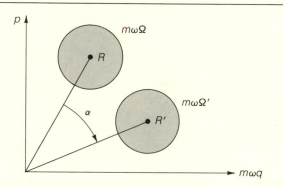

Figure 3.6 Motion of the phase volume along the phase orbits for a one-dimensional harmonic oscillator.

III THE STATISTICAL ENSEMBLE

4. Distribution Function and Probability Density(*)

Given the form of the hamiltonian for a specific system, the integration of Hamilton's equations allows one in principle to uniquely determine all the dynamical variables at any time, provided that sufficient data are available to yield their values at a fixed time t_0. In dealing with macroscopic systems of very many degrees of freedom, however, there are normally far too few data furnished by actual observation to satisfy this requirement, so that further information can be based only upon statements of probability.

To include all possibilities, it is necessary to consider any set (x_k) of dynamical variables $(k = 1, 2, \ldots, 2n)$, for the set (p_i, q_i) of momenta and coordinates. One may thus imagine an arbitrarily large number ν of identically reproduced systems, each pertaining to a separate set of these variables. Such a manifold is called an *ensemble*, with the understanding that any one of its members shall be required to be a possible representative of the individual system under investigation.

The ensemble can be characterized by ν different points in the phase space such that each of them represents a separate member with the corresponding set (x_k) given by the instantaneous values of the dynamical variables. Considering the points within a given region, and denoting by $\Delta\nu$ their number at a certain time, one then has by definition

$$\Delta P = \frac{\Delta\nu}{\nu} \tag{4.1}$$

for the probability of finding at that time the dynamical variables of the system under investigation within the range of values indicated by this region (see Figure 4.1).

In the limit of an infinitesimal region with a range of variables between x_k and $x_k + dx_k$, one is thus led to the differential probability

$$dP = \frac{d\nu}{\nu} \tag{4.2}$$

By choosing ν sufficiently large, one can further assume the representative points to be so densely located that it is allowed, in effect, to regard their number $d\nu$ within any such region as a continuously varying quantity. With the volume element in phase space defined by (see Sec. 3)

$$d\lambda = \prod_{k=1}^{2n} dx_k = \prod_{i=1}^{n} dp_i dq_i \tag{4.3}$$

Whereas Eq.(4.5) connects the distribution function with the number of members assumed to constitute the ensemble, the probability density is independent of this number. Instead, one has from Eq.(4.8)

$$\int \rho(x)d\lambda \ = \ 1 \qquad (4.9)$$

so that the integrated value of all the differential probabilities dP has the required property to be equal to unity. Equation (4.9) thus expresses the obvious certainty that the variables of the system under investigation, whatever their actual value, pertain to a set represented by one of the points in the phase space.

Whereas D and ρ have been introduced as functions of a definite set (x_k) of variables, it shall be noted that their significance remains unchanged under a canonical transformation that preserves the volume element $d\lambda$ (this is proven in Appendix B).

5. Mean Values(*)

Once known, the distribution function $D(x)$ contains the most general description of an ensemble, and the corresponding probability density $\rho(x)$ allows the answer to any question of probability. One may thus ask for the probability P_R of finding the dynamical variables within a range of their values, represented by a given region R of phase space. From the alternative differential probabilities dP of Eq.(4.6), referring to any volume element $d\lambda$ within that region, it follows that

$$P_R \ = \ \int_{(R)} \rho(x)d\lambda \qquad (5.1)$$

with the integral extended over the entire region R.

Although the set $(x) \ \equiv \ (x_k)$ with $k \ = \ 1, 2, \ldots, 2n$ of dynamical variables (p_i, q_i) accounts in full detail for the state of a system, there are certain special quantities, such as the energy, given as functions of these variables, which are usually of far more direct significance. One can then ask for the probability $P(\phi_1, \phi_2)$ of a function $\phi(x)$ to be found within an interval between two given values ϕ_1 and ϕ_2. According to Eq.(5.1), this probability is given by

$$P(\phi_1, \phi_2) \ = \ \int_{(R_{1,2})} \rho(x)d\lambda \qquad (5.2)$$

where the integral extends over the region $R_{1,2}$, for which

$$\phi_1 \le \phi(x) \le \phi_2 \qquad (5.3)$$

The knowledge of this probability for an arbitrary interval permits one to obtain any statistical information about the values of ϕ to be expected. There are simpler means, however, to arrive at significant, albeit more qualitative, conclusions.

For this purpose consider the mean value

$$\bar{\phi} \equiv \int \phi(x)\rho(x)d\lambda \qquad (5.4)$$

of the quanitity ϕ and the mean square of its deviation

$$\Delta\phi = \phi - \bar{\phi} \qquad (5.5)$$

from this value, defined by

$$\overline{(\Delta\phi)^2} \equiv \int [\phi(x) - \bar{\phi}]^2 \rho(x)d\lambda \qquad (5.6)$$

both being weighted by the probability density with the integrals extended over the entire phase space. By expansion of the square in the integrand of Eq.(5.6), and in view of Eqs.(4.9) and (5.4), one obtains the equivalent expression

$$\overline{(\Delta\phi)^2} = \overline{(\phi^2)} - (\bar{\phi})^2 \qquad (5.7)$$

where

$$\overline{(\phi^2)} \equiv \int \phi^2(x)\rho(x)d\lambda \qquad (5.8)$$

represents the mean square of ϕ . It is customary, further, to use for the positive square root of $\overline{(\Delta\phi)^2}$ the notation

$$\sqrt{\overline{(\Delta\phi)^2}} \equiv (\Delta\phi)_{\text{rms}} \qquad (5.9)$$

and to call it the *root-mean-square deviation.*

Besides the deviation of ϕ from the mean value, one could also consider the deviation

$$\Delta\phi^* = \phi - \phi^* \qquad (5.10)$$

from an arbitrary given value ϕ^*. The special significance of the mean value $\bar{\phi}$ arises from the fact that it represents an optimal choice in the sense that

$$\overline{(\Delta\phi^*)} = \int [\phi(x) - \phi^*]\rho(x)d\lambda \qquad (5.11)$$

is seen to vanish and

$$\overline{(\Delta\phi^*)^2} = \int [\phi(x) - \phi^*]^2 \rho(x)d\lambda \qquad (5.12)$$

to reach its minimum by assigning to ϕ^* the particular value $\bar{\phi}$.

In order to judge the significance of the mean value $\bar{\phi}$, it is important to obtain some information about the magnitude of the deviation $\Delta\phi$ to be expected for the system under actual observation. Whereas the mean deviation $\overline{(\Delta\phi)} = \int [\phi(x) - \bar{\phi}]\rho(x)d\lambda$ is identically zero and therefore unsuitable for this purpose, the value of $(\Delta\phi)_{\mathrm{rms}}$, defined by Eqs.(5.7) and (5.9), will be shown to indicate the range within which deviations of the quantity ϕ from its mean value are most likely to occur.

Since the integrand on the right side of Eq.(5.6) is positive, it follows in particular that $\overline{(\Delta\phi)^2}$ and hence $(\Delta\phi)_{\mathrm{rms}}$ does not vanish unless $\rho(x)$ has the singular property to differ from zero only on the surface of the phase space where $\phi(x) = \bar{\phi}$ and here to become infinite in such a way as to still satisfy the requirement of Eq.(4.9). A vanishing result for $(\Delta\phi)_{\mathrm{rms}}$ therefore has the significance that ϕ will be found with certainty to have the value $\bar{\phi}$.

A finite value of $(\Delta\phi)_{\mathrm{rms}}$ does not lead to an equally unambiguous conclusion. It can be interpreted, however, by choosing an arbitrary positive number C and considering the region R in the phase space for which the absolute magnitude of $\Delta\phi$ from Eq.(5.5) is sufficiently large to satisfy the condition

$$| \phi(x) - \bar{\phi} | \geq C(\Delta\phi)_{\mathrm{rms}} \tag{5.13}$$

From the square of this relation and the use of the definition in Eq.(5.9), it follows upon multiplication with $\rho(x)$ and integration over R on both sides that

$$\int_{(R)} [\phi(x) - \bar{\phi}]^2 \rho(x)d\lambda \geq C^2\overline{(\Delta\phi)^2} \int_{(R)} \rho(x)d\lambda \tag{5.14}$$

On the other hand, the integral on the right side of Eq.(5.6) can be separated into one extending over the region R, and another over the remaining region of the phase space. Upon omission of the latter, one obtains therefore the second inequality

$$\overline{(\Delta\phi)^2} \geq \int_{(R)} [\phi(x) - \bar{\phi}]^2 \rho(x)d\lambda \tag{5.15}$$

A combination with Eq.(5.14) that implies

$$\overline{(\Delta\phi)^2} \geq C^2\overline{(\Delta\phi)^2} \int_{(R)} \rho(x)d\lambda \tag{5.16}$$

Since $\overline{(\Delta\phi)^2}$ is here assumed to be finite, this leads to the conclusion

$$P_R \leq 1/C^2 \tag{5.17}$$

where P_R is given by Eq. (5.1). With the present definition of the region R, it is evident that P_R represents the probability for $|\Delta\phi| = |\phi(x) - \bar{\phi}|$ of satisfying the condition of Eq.(5.13).

It should be noted that P_R may actually be very much smaller than the upper limit demanded by the relation of Eq.(5.17). In the typical example of a gaussian dependence of the differential probability to find ϕ within an interval $d\phi$, it follows in fact for large values of C that $P_R \cong [\exp(-C^2/2)]/C \ll 1/C^2$. Another example arises if the function $\phi(x)$ cannot exceed a finite value. With $\bar{\phi}$ and $(\Delta\phi)_{\rm rms}$ likewise finite, it is then impossible to satisfy the condition of Eq.(5.13) for C chosen larger than a certain value $|\phi(x) - \bar{\phi}|_{\max}/(\Delta\phi)_{\rm rms}$, a fact expressed by the equivalent result $P_R = 0$.

The significance of the general conclusion reached in Eq.(5.17) becomes more transparent by letting $C \gg 1$. Then

$$P_R \ll 1 \tag{5.18}$$

when the condition of Eq.(5.13) satisfied by the region R is

$$|\phi(x) - \bar{\phi}| \gg (\Delta\phi)_{\rm rms} \tag{5.19}$$

This states that it is highly improbable to find deviations $\Delta\phi$ with an absolute value much larger than their root-mean-square $(\Delta\phi)_{\rm rms}$. Conversely, $(\Delta\phi)_{\rm rms}$ is therefore seen to represent the order of magnitude of the range around $\bar{\phi}$ within which ϕ is almost certain to be found.

The preceding considerations are of particular importance for the characterization of thermodynamical variables, used in the description of macroscopic systems. One relies here upon the effectively unambiguous value of certain quantities without the information being available that would account for the detailed state of the system. Such quantities are to be statistically interpreted as referring to functions $\phi(x)$ of the dynamical variables for which the probability $P(\phi_1, \phi_2)$ of Eq.(5.2) has the property that it becomes close to unity even for a relatively small interval between ϕ_1 and ϕ_2. As a more convenient criterion, these functions can be required to satisfy the condition

$$(\Delta\phi)_{\rm rms}/|\bar{\phi}| \ll 1 \tag{5.20}$$

In view of the previously established significance of $(\Delta\phi)_{\rm rms}$, this relation allows in fact a practically unique assignment of ϕ, given by its mean value

$\bar{\phi}$, insofar as the actual value can be expected with high probability, to differ by no more than a relatively small amount.

The very use of thermodynamical variables implies that such an assignment is justified for the quantities and systems with which one deals in thermodynamics and indicates that the validity of Eq.(5.20) can here be safely assumed for reasons peculiar to the concern with macroscopic features. In order to perform a test, these features therefore have to be incorporated in the choice of the quantities considered and of the probability density ρ required for the evaluation of mean values according to Eqs.(5.4), (5.6), and (5.8). The remainder of this section shall serve to illustrate typical aspects of such a test.

Additive Quantities. With the molecular constitution of macroscopic bodies as an important representative, it is of interest to consider the case of a system composed of a large number N of subsystems. The quantities under consideration will here be chosen to have the *additive property*, expressed by

$$\phi = \sum_s \phi_s \tag{5.21}$$

The summation over s extends from 1 to N, and ϕ_s is a function of the dynamical variables (x_s) that pertain to the subsystem s. With the calculation of mean values, one has then

$$\bar{\phi} = \sum_s \bar{\phi}_s \tag{5.22}$$

and from Eq.(5.6), upon expanding the square of the sums in Eqs.(5.21) and (5.22)

$$\overline{(\Delta\phi)^2} = \sum_s [\overline{(\phi_s^2)} - (\bar{\phi}_s)^2] + \sum_{s \neq s'} [\overline{(\phi_s \phi_{s'})} - (\bar{\phi}_s \bar{\phi}_{s'})] \tag{5.23}$$

As a choice of the probability density ρ in the present example, it shall be assumed that one deals with a product

$$\rho = \prod_s \rho_s \tag{5.24}$$

where ρ_s is likewise a function of the variables (x_s) such that

$$\int \rho_s d\lambda_s = 1 \tag{5.25}$$

and where the integral extends over the phase space of these variables with the volume element denoted by $d\lambda_s$. In view of Eq.(4.3), the volume element $d\lambda$ of the entire phase space is the product of the volume elements $d\lambda_s$, so that as a consequence of Eq.(5.24)

$$\rho d\lambda = \prod_s \rho_s d\lambda_s \qquad (5.26)$$

With

$$dP_s = \rho_s d\lambda_s \qquad (5.27)$$

representing the differential probability of finding the variables of the subsystem s within the volume element $d\lambda_s$, it follows from Eqs.(5.26) and (5.27) that the total differential probability dP of Eq.(4.6) is given by

$$dP = \prod_s dP_s \qquad (5.28)$$

Since this relation characterizes the joint result of independent probabilities, the equivalent form of Eq.(5.24) is thus seen to have the significance that the subsystems are here assumed to be *statistically uncorrelated*.

In view of Eq.(5.26), the integrals over the entire phase space, required to obtain mean values, appear for the terms on the right side of Eqs.(5.22) and (5.23) as products of integrals over the phase space of individual subsystems. One has therefore

$$\bar{\phi}_s = \int \phi_s \rho_s d\lambda_s \qquad (5.29)$$

since Eq.(5.25) yields unity for all other factors, and for the same reason

$$\overline{(\phi_s^2)} = \int \phi_s^2 \rho_s d\lambda_s \qquad (5.30)$$

Similarly, one obtains for $s \neq s'$

$$\overline{(\phi_s \phi_{s'})} = \left(\int \phi_s \rho_s d\lambda_s \right) \left(\int \phi_{s'} \rho_{s'} d\lambda_{s'} \right) \qquad (5.31)$$

or from Eq.(5.29)

$$\overline{(\phi_s \phi_{s'})} = \bar{\phi}_s \bar{\phi}_{s'} \qquad (5.32)$$

so that the second sum of Eq.(5.23) is seen to vanish. In analogy to Eqs.(5.6) and (5.7), the term

$$\overline{(\Delta \phi_s)^2} = \overline{(\phi_s^2)} - (\bar{\phi}_s)^2 \qquad (5.33)$$

in the remaining sum represents the mean square deviation $\overline{(\phi_s - \bar{\phi}_s)^2}$ of ϕ_s. Under the present assumption of uncorrelated subsystems, it thus follows that the mean square deviation of an additive quantity likewise has the additive property and is given by

$$\overline{(\Delta\phi)^2} = \sum_s \overline{(\Delta\phi_s)^2} \tag{5.34}$$

For simplicity, it shall finally be assumed that there is no distinction among the subsystems so that the dependence of both ϕ_s and ρ_s on the set of variables (x_s) is the same for every subsystem. One is thus led to the common mean values

$$\bar{\phi}_s \equiv \bar{\varphi} \tag{5.35}$$

in Eq.(5.22) as well as

$$\overline{(\Delta\phi_s)^2} \equiv \overline{(\Delta\varphi)^2} \tag{5.36}$$

in Eq.(5.34) for all terms of the sum over s from 1 to N, and hence to

$$\bar{\phi} = N\bar{\varphi} \tag{5.37}$$

$$\overline{(\Delta\phi)^2} = N\overline{(\Delta\varphi)^2} \tag{5.38}$$

From the definition of $(\Delta\phi)_{\text{rms}}$ by Eq.(5.9), and using the analogous notation

$$(\Delta\varphi)_{\text{rms}} \equiv \sqrt{\overline{(\Delta\varphi)^2}} \tag{5.39}$$

it then follows that

$$\frac{(\Delta\phi)_{\text{rms}}}{|\bar{\phi}|} = \frac{1}{\sqrt{N}} \frac{(\Delta\varphi)_{\text{rms}}}{|\bar{\varphi}|} \tag{5.40}$$

A subsystem may well have only a few degrees of freedom and a ratio $(\Delta\varphi)_{\text{rms}}/|\bar{\varphi}|$ comparable to, or even larger than, unity. Irrespective of this ratio it is seen, however, that the condition of Eq.(5.20) is the more closely fulfilled the larger the number N of subsystems. For sufficiently large N it is therefore justified, in the case considered, to assign to the quantity ϕ the average value $\bar{\phi}$, since any deviation from this value can then be confidently expected to call for a negligible correction.

Among examples, one may think of a homogeneous solid as being built up of many equal pieces to represent the subsystems, with ϕ and φ respectively chosen as the energy of the whole solid and of one of the pieces. It is true that the interaction between neighboring pieces across their common contact surface violates the additivity assumed by Eq.(5.21) and provides a mechanism for correlations, assumed by Eq.(5.24) to be absent. Both assumptions are acceptable, however, provided that the linear dimensions of the pieces are chosen large enough for the molecules in their interior to be far more numerous than those on the contact surfaces, so that effects of interactions across the latter can be neglected. One is then allowed to apply Eq.(5.40) and to conclude that there is a unique value for the energy of a sufficiently large solid.

In dealing with the molecular constitution of material systems, it is appropriate to regard the molecules as the subsytems considered in the preceding analysis. One can discuss, for example, molecular distributions specifically directed towards the case of principal importance in the kinetic theory of gases, where the density is sufficiently low for the interaction between molecules only to manifest itself in relatively short interruptions of their free motion during a collision. The system may thus be considered, in effect, to be an ideal gas with an energy given by the total kinetic energy of all its molecules. This important application is discussed in more detail in Appendix C.

6. Time Dependence of the Phase Density(*)

As defined by Eq.(4.4), the distribution function D determines the number $d\nu$ of members of the ensemble which at a given time are represented by the points within a given volume element $d\lambda$ of the phase space. Except for special circumstances, $d\nu$ will be found to change in time due to the difference between the number of points entering and of those leaving the volume element during a time interval. It is thus expected that one deals generally with a dependence of the distribution function not only on the dynamical variables (x) but also on the time t, to be indicated by writing

$$D \;=\; D(x,t) \tag{6.1}$$

As long as no decision has been made about the type of information available, expressed in the choice of this function, the form of D at a given time remains open. Having once chosen a definite form $D(x,t_0)$ at a given time t_0, however, this freedom no longer exists at any other time, since the evolution of every set of dynamical variables, given at the time t_0, is uniquely determined by integration of Hamilton's equations.

In order to arrive at the ensuing equation for the distribution function, let

$$d\nu_0 \;=\; D(x_0,t_0)d\lambda_0 \tag{6.2}$$

indicate the number of members of the ensemble that lie within the volume element $d\lambda_0$ located at the point in phase space characterized by the set of variables (x_0) at the time t_0. If the representative point of any one of these members is followed along its phase orbit up to the time t, then the set (x_0) goes over into the set $[x(t)]$ and the volume element $d\lambda_0$ goes into the volume element $d\lambda_t$. The situation is illustrated in Figure 6.1. By definition, the number of members of the ensemble in the volume $d\lambda_t$ at the point x at time t is given by

$$d\nu_t = D[x(t), t]d\lambda_t \qquad (6.3)$$

Since each representative point found within $d\lambda_0$ will also be found within $d\lambda_t$ and vice versa, *their number is the same in both*, so that

$$d\nu_0 = d\nu_t \qquad (6.4)$$

Furthermore, according to Liouville's Theorem [Eq.(3.23)] one has

$$d\lambda_0 = d\lambda_t \qquad (6.5)$$

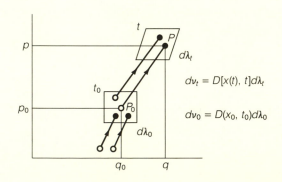

Figure 6.1   Schematic representation of the time evolution of the representative points in phase space of the members of an ensemble. Here (x) stands for the variable set (p, q).

and hence from Eqs.(6.2) and (6.3) the distribution function must satisfy the following relation

$$D[x(t), t] = D[x_0, t_0] \tag{6.6}$$

We conclude that *the distribution function is unchanged along a phase orbit* (Fig. 6.1).

Now the relation (6.6) must hold at all times, and since the right side does not depend on t, it follows that

$$\frac{d}{dt} D[x(t), t] = 0 \tag{6.7}$$

If we recall that the set (x) is simply a shorthand for the full set of dynamical variables (p_i, q_i) with $i = 1, 2, \ldots, n$, then Eq.(6.7) can be written in detail as

$$\sum_{i=1}^{n} \left(\frac{\partial D}{\partial p_i} \dot{p}_i + \frac{\partial D}{\partial q_i} \dot{q}_i \right) + \frac{\partial D}{\partial t} = 0 \tag{6.8}$$

Upon the insertion of Hamilton's equations

$$\begin{aligned}
\dot{p}_i &= -\frac{\partial H}{\partial q_i} \qquad ; \; i = 1, 2 \ldots, n \\
\dot{q}_i &= \frac{\partial H}{\partial p_i}
\end{aligned} \tag{6.9}$$

one obtains

$$\frac{\partial D}{\partial t} = -\sum_{i=1}^{n} \left(\frac{\partial D}{\partial q_i} \frac{\partial H}{\partial p_i} - \frac{\partial D}{\partial p_i} \frac{\partial H}{\partial q_i} \right) \tag{6.10}$$

as the differential equation to be satisfied by the function D of Eq.(6.1). It gives the time-rate-of-change of the density function D at a given point in phase space, since the other members of the variable set (p_i, q_i, t) are to be held fixed in evaluating the partial derivatives in this expression. Since Eq.(6.10) is of first order in time, it indeed permits a free choice in the dependence of D on the dynamical variables at the time t_0, but integration of this result then leads to a uniquely determined form of D at any other time t.

The importance of Eq.(6.10) merits another derivation, based upon a more intuitive approach. In fact, the points in the phase space, representing the members of the ensemble, can be regarded as analogous to the particles in the three-dimensional space that constitute a fluid in motion. If n denotes the number of these particles per unit volume and **v** the local drift velocity vector, then the equation of continuity for the fluid states that

$$\frac{\partial n}{\partial t} + \text{div } (\mathbf{v}n) = 0 \tag{6.11}$$

In analogy, with n replaced by D and the components of **v** by dx_k/dt, one obtains the relation

$$\frac{\partial D}{\partial t} + \sum_{k=1}^{2n} \frac{\partial}{\partial x_k}\left(\frac{dx_k}{dt}D\right) = 0 \tag{6.12}$$

Indeed, integrated over a closed region of phase space, and by the analogue of Gauss's Theorem, this relation expresses the fact that the rate of change in the number of representative points within that region is given by the difference between the number per unit time of those that enter and leave the enclosing surface. In detail, Eq.(6.12) states

$$\frac{\partial D}{\partial t} + \sum_{i=1}^{n}\left[\frac{\partial}{\partial p_i}(\dot{p}_iD) + \frac{\partial}{\partial q_i}(\dot{q}_iD)\right] = 0 \tag{6.13}$$

and the insertion of Hamilton's Eqs.(6.9) leads to

$$\frac{\partial D}{\partial t} + \sum_{i=1}^{n}\left[-\frac{\partial}{\partial p_i}\left(D\frac{\partial H}{\partial q_i}\right) + \frac{\partial}{\partial q_i}\left(D\frac{\partial H}{\partial p_i}\right)\right] = 0 \tag{6.14}$$

Since interchange of the order of partial differentiation is always permissible, the mixed derivatives $\partial^2 H/\partial p_i \partial q_i$ cancel in this expression, and one obtains

$$\frac{\partial D}{\partial t} + \sum_{i=1}^{n}\left(\frac{\partial D}{\partial q_i}\frac{\partial H}{\partial p_i} - \frac{\partial D}{\partial p_i}\frac{\partial H}{\partial q_i}\right) = 0 \tag{6.15}$$

which is precisely Eq. (6.10).

The previous derivation of Eq.(6.10), besides using Hamilton's equations, was based on Liouville's Theorem. It may be noted that since the derivation just presented implies

$$\frac{d}{dt}D[x(t), t] = 0 \tag{6.16}$$

and hence allows one to *conclude* Eq.(6.5), it can be regarded as an alternate proof of Liouville's Theorem. The connection is somewhat indirect, however, because the concept of an ensemble, important as it

is for statistical considerations, enters in this alternate derivation merely as an auxiliary, introduced to prove a theorem of pure mechanics.

We proceed to analyze Eq.(6.10) in more detail. For this purpose, consider first two arbitrary functions $A(p_i, q_i, t)$ and $B(p_i, q_i, t)$. The *Poisson Bracket* of these two functions is defined by

$$[A, B]_{P.B.} \equiv \sum_{i=1}^{n} \left(\frac{\partial A}{\partial q_i} \frac{\partial B}{\partial p_i} - \frac{\partial A}{\partial p_i} \frac{\partial B}{\partial q_i} \right) \tag{6.17}$$

$$= -[B, A]_{P.B.}$$

Now the total time derivative of any function $\phi(p_i, q_i, t)$ is evidently given by

$$\frac{d\phi}{dt} = \sum_{i=1}^{n} \left(\frac{\partial \phi}{\partial p_i} \dot{p}_i + \frac{\partial \phi}{\partial q_i} \dot{q}_i \right) + \frac{\partial \phi}{\partial t} \tag{6.18}$$

and the use of Hamilton's Eqs.(6.9) leads to

$$\frac{d\phi}{dt} = \frac{\partial \phi}{\partial t} + \sum_{i=1}^{n} \left(\frac{\partial \phi}{\partial q_i} \frac{\partial H}{\partial p_i} - \frac{\partial \phi}{\partial p_i} \frac{\partial H}{\partial q_i} \right) \tag{6.19}$$

or

$$\frac{d\phi}{dt} = \frac{\partial \phi}{\partial t} - [H, \phi]_{P.B.} \tag{6.20}$$

This Poisson Bracket relation provides a formulation of classical mechanics that is fully equivalent to Hamilton's equations and hence to Newton's Laws. We have shown that it follows from Hamilton's equations. Conversely, Hamilton's equations are immediately derived from Eq.(6.20), since the defining relation (6.17) implies that

$$[H, p_i]_{P.B.} = \frac{\partial H}{\partial q_i} \qquad ; \; i = 1, 2 \ldots, n$$
$$[H, q_i]_{P.B.} = -\frac{\partial H}{\partial p_i} \tag{6.21}$$

and application of Eq.(6.20) to $\phi(p_i, q_i, t) = p_i$ and q_i in turn then gives

$$\frac{dp_i}{dt} = -\frac{\partial H}{\partial q_i} \qquad ; \; i = 1, 2 \ldots, n$$
$$\frac{dq_i}{dt} = \frac{\partial H}{\partial p_i} \tag{6.22}$$

which are indeed Hamilton's equations. Furthermore, with $\phi \equiv H(p_i, q_i, t)$ one has

$$\frac{dH}{dt} = \frac{\partial H}{\partial t} - [H, H]_{P.B.} = \frac{\partial H}{\partial t} \tag{6.23}$$

which is precisely Eq.(1.60). The elementary Poisson Brackets follow immediately from the defining relation (6.17)

$$[p_i, q_j]_{P.B.} = -\delta_{ij}$$
$$[p_i, p_j]_{P.B.} = [q_i, q_j]_{P.B.} = 0 \tag{6.24}$$

Starting from the description of a given mechanical system by the set (p_i, q_i) of variables with $i = 1, 2 \ldots, n$, one may go over to a new set (p'_k, q'_k) with $k = 1, 2 \ldots, n$ defined as functions of the former through

$$p'_k = p'_k(p_i, q_i) \qquad ; \; k = 1, 2 \ldots, n$$
$$q'_k = q'_k(p_i, q_i) \tag{6.25}$$

or the inverse relations

$$p_i = p_i(p'_k, q'_k) \qquad ; \; i = 1, 2 \ldots, n$$
$$q_i = q_i(p'_k, q'_k) \tag{6.26}$$

as discussed in Sec. 1. An alternative definition of the *canonical transformations* discussed in that section, under which the formulation of classical mechanics remains unchanged, is that Poisson Brackets should be preserved. As shown in Appendix A, it is sufficient for this purpose that the relations

$$[p'_k, q'_l]_{P.B.} = -\delta_{kl}$$
$$[p'_k, p'_l]_{P.B.} = [q'_k, q'_l]_{P.B.} = 0 \tag{6.27}$$

be preserved. Then, for example, if

$$H = H[p_i(p'_k, q'_k), q_i(p'_k, q'_k), t] \equiv H'(p'_k, q'_k, t) \tag{6.28}$$

[c.f. Eq. (1.43)] with a smiliar relation for ϕ, it will again be true that

$$\frac{d\phi'}{dt} = \frac{\partial \phi'}{\partial t} - [H', \phi']_{P.B.'} \tag{6.29}$$

where the Poisson Bracket is now computed with respect to the variable set (p'_k, q'_k, t)

$$[A', B']_{P.B.'} \equiv \sum_{k=1}^{n} \left(\frac{\partial A'}{\partial q'_k} \frac{\partial B'}{\partial p'_k} - \frac{\partial A'}{\partial p'_k} \frac{\partial B'}{\partial q'_k} \right) \tag{6.30}$$

A repetition of the arguments in Eqs.(6.21)-(6.22) shows that Hamilton's equations are preserved under these transformations

$$
\begin{aligned}
\frac{dp'_k}{dt} &= -\frac{\partial H'}{\partial q'_k} \qquad ; \; k = 1, 2 \ldots, n \\
\frac{dq'_k}{dt} &= \frac{\partial H'}{\partial p'_k}
\end{aligned}
\tag{6.31}
$$

exactly as in Eqs.(1.42).

The definition of the Poisson Bracket in Eq.(6.17) with $A = H$ and $B = D$ allows the basic result for the time development of the distribution function in Eq.(6.10) to be written compactly as

$$
\frac{\partial D}{\partial t} - [H, D]_{P.B.} = 0
\tag{6.32}
$$

and as just discussed, this relation is invariant under a canonical transformation of the dynamical variables.

According to the definition of Eq.(4.8), the probability density ρ differs from the distribution function D merely by the constant factor $\nu = \int D(x)d\lambda$. The time dependence of D indicated in Eq.(6.1) is therefore reflected in that of ρ

$$
\rho = \rho(x, t)
\tag{6.33}
$$

and the differential equation (6.10) or (6.32) for the function $D(x, t)$ retains its form for the function $\rho(x, t)$, so that one has likewise

$$
\frac{\partial \rho}{\partial t} + \sum_{i=1}^{n} \left(\frac{\partial \rho}{\partial q_i} \frac{\partial H}{\partial p_i} - \frac{\partial \rho}{\partial p_i} \frac{\partial H}{\partial q_i} \right) = 0
\tag{6.34}
$$

or

$$
\frac{\partial \rho}{\partial t} - [H, \rho]_{P.B.} = 0
\tag{6.35}
$$

In contrast to the fact that a multiplication of D by a constant factor merely affects the arbitrary number ν of members in the ensemble, Eq. (4.9) requires ρ to satisfy the condition

$$
\int \rho(x, t)d\lambda = 1
\tag{6.36}
$$

at any time.

IV THERMAL EQUILIBRIUM AND THE CANONICAL DISTRIBUTION

7. Stationary Mean Values(*)

Consider an arbitrary function $Q = Q(p_i, q_i)$ of the momenta and coordinates. This function evidently develops in time according to

$$\dot{Q} = \frac{dQ}{dt} = \sum_{i=1}^{n} \left(\frac{\partial Q}{\partial p_i} \dot{p}_i + \frac{\partial Q}{\partial q_i} \dot{q}_i \right) \tag{7.1}$$

The *ensemble average*, or *mean value*, or Q, is given by Eq.(5.4) as

$$\bar{Q} \equiv \int Q(p_i, q_i) \rho(q_i, p_i, t) d\lambda \tag{7.2}$$

where $\rho(p_i, q_i, t)$ is the probability density of Eq.(4.8), and from Eq.(4.3)

$$d\lambda \equiv \prod_{i=1}^{n} dp_i dq_i \tag{7.3}$$

In our shorthand where $(x) \equiv (x_k)$ with $k = 1, 2 \ldots, 2n$ stands for the set (p_i, q_i) with $i = 1, 2 \ldots, n$, these equations were previously abbreviated as

$$\bar{Q} = \int Q(x) \rho(x, t) d\lambda \tag{7.4}$$

with

$$d\lambda = \prod_{k=1}^{2n} dx_k \tag{7.5}$$

Now, although Q does not depend *explicitly* on t, the mean value of $\bar{Q} = \bar{Q}(t)$ depends on t through the time dependence of the probability density $\rho(x, t)$ [Eq.(6.33)]. It follows that the time derivative of $\bar{Q}(t)$ is given by

$$\frac{d\bar{Q}}{dt} = \int Q(p_i, q_i) \frac{\partial}{\partial t} \rho(p_i, q_i, t) d\lambda \tag{7.6}$$

Note that it is the partial time derivative that occurs inside the integral. The set (p, q_i) now merely represents the phase space integration variables of Eq.(7.3), and they are kept constant while the time derivative of the integral is evaluated. We have previously demonstrated in Eq.(6.7) and (6.34) that the *total* time derivative of the probability density $\rho(p_i, q_i, t)$ must vanish

$$\frac{d\rho}{dt} = \frac{\partial \rho}{\partial t} + \sum_{i=1}^{n}\left(\frac{\partial \rho}{\partial q_i}\dot{q}_i + \frac{\partial \rho}{\partial p_i}\dot{p}_i\right) = 0 \tag{7.7}$$

Equation (7.6) can thus be rewritten as

$$\frac{d\bar{Q}}{dt} = -\int Q(p_i, q_i)\left[\sum_{i=1}^{n}\left(\frac{\partial \rho}{\partial q_i}\dot{q}_i + \frac{\partial \rho}{\partial p_i}\dot{p}_i\right)\right]d\lambda \tag{7.8}$$

Since ρ is positive and $\int \rho d\lambda = 1$, the convergence of the integral requires that $\rho \to 0$ *for* $|p_i| \to \infty$ *or* $|q_i| \to \infty$. By partial integration with respect to these variable one has therefore

$$\frac{d\bar{Q}}{dt} = \int \rho\left\{\sum_{i=1}^{n}\left[\frac{\partial}{\partial p_i}\Big(Q(p_i, q_i)\dot{p}_i\Big) + \frac{\partial}{\partial q_i}\Big(Q(p_i, q_i)\dot{q}_i\Big)\right]\right\}d\lambda \tag{7.9}$$

The dynamics of the variables (p_i, q_i) are given by Hamilton's Eqs.(6.9). It follows that

$$\frac{\partial}{\partial p_i}\dot{p}_i + \frac{\partial}{\partial q_i}\dot{q}_i = \frac{\partial}{\partial p_i}\left(\frac{-\partial H}{\partial q_i}\right) + \frac{\partial}{\partial q_i}\left(\frac{\partial H}{\partial p_i}\right) = 0 \tag{7.10}$$

since the order of partial differentiation can be interchanged. Equation (7.9) thus simplifies to

$$\frac{d\bar{Q}}{dt} = \int \rho\left[\sum_{i=1}^{n}\left(\frac{\partial Q}{\partial p_i}\dot{p}_i + \frac{\partial Q}{\partial q_i}\dot{q}_i\right)\right]d\lambda \tag{7.11}$$

$$= \int \rho(p_i, q_i, t)\frac{d}{dt}Q(p_i, q_i)d\lambda \tag{7.12}$$

where Eq.(7.1) has been used in the second line. The righthand side of this result is recognized from Eq.(5.4) to be the mean value of the quantity dQ/dt, and thus we have

$$\frac{d\bar{Q}}{dt} = \overline{\left(\frac{dQ}{dt}\right)} \tag{7.13}$$

The analysis in Eqs.(6.17)-(6.20) allows Eqs.(7.11)-(7.13) to be reexpressed in the equivalent forms

$$\frac{d\bar{Q}}{dt} = -\int [H, Q]_{P.B.}\, \rho\, d\lambda$$

$$= \int \frac{dQ}{dt}\, \rho\, d\lambda \tag{7.14}$$

$$= \overline{\left(\frac{dQ}{dt}\right)}$$

In words, this result states that the time-derivative of the mean value is the mean value of the time derivative. The analogous result in quantum mechanics is known as Ehrenfest's Theorem (see Sec. 20).

Conditions of Equilibrium. The laws of thermodynamics are primarily based upon the consideration of equilibrium, that is, of those states of a macroscopic system in which the thermodynamical variables, used for its description, remain stationary. Among the transitions from one such state to another, one therefore has to pay special attention to those, occurring in *reversible* processes, where the change of external conditions is sufficiently gradual for the system to pass in effect through a continuous sequence of equilibria. With the more recent branch of "irreversible thermodynamics" left aside, the goal will be here to establish the connection of statistical mechanics to classical "reversible" thermodynamics.

In contrast to mechanical equilibrium, the existence of thermal equilibrium does not demand the strict absence of any variation in time. Consider, for example, the number of molecules in a certain space-time region of a gas or liquid; not equally many of them are likely to leave and enter that region during a given time interval. One therefore has to expect that this number will fluctuate over a range that can be relatively more significant, the smaller the region. Similarly, the impact of molecules causes a fluctuating force to act on a surface that is manifested in the Brownian motion of small particles suspended in a liquid.

As pointed out in Sec. 5, the use of thermodynamical variables implies the reliance upon an essentially unique value, given by the mean value $\bar{\phi}$ of the corresponding quantities ϕ. To the extent to which this reliance is justified, *thermodynamic equilibrium* may thus be characterized by the requirement that the mean value $\bar{\phi}$ remain constant in time

$$\frac{d\bar{\phi}}{dt} = 0 \tag{7.15}$$

Consider, then, the mean value of a dynamical quantity $Q(p_i, q_i)$

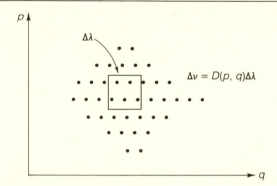

Figure 4.1 Illustration of the occupied points in phase space corresponding to the members of an *ensemble* of systems. There are $\Delta\nu$ members of
the ensemble in the volume element $\Delta\lambda$.

one may then write

$$d\nu \;=\; D(x)d\lambda \tag{4.4}$$

where $D(x)$ is a function of the variables x_k, called the *distribution
function*. By integration over the whole of phase space, it follows that

$$\nu \;=\; \int D(x)d\lambda \tag{4.5}$$

and hence, from Eq.(4.2), that

$$dP \;=\; \frac{D(x)d\lambda}{\int D(x)d\lambda} \tag{4.6}$$

is the probability of finding the set of variables (x_k) within the infinitesimal region of volume $d\lambda$.

As another form of Eq.(4.6), one can write

$$dP \;=\; \rho(x)d\lambda \tag{4.7}$$

where

$$\rho(x) \;=\; \frac{D(x)}{\int D(x)d\lambda} \tag{4.8}$$

shall be called the *probability density*.

$$\bar{Q} = \bar{Q}(t) = \int Q(p_i, q_i)\rho(p_i, q_i, t)d\lambda \qquad (7.16)$$

The condition of thermodynamic equilibrium in Eq.(7.15) states that

$$\frac{d\bar{Q}}{dt} = \int Q(p_i, q_i)\frac{\partial}{\partial t}\rho(p_i, q_i, t)d\lambda = 0 \qquad (7.17)$$

It shall be assumed for the present development that this relation holds rigorously irrespective of the quantity $Q(p_i, q_i)$ to be considered. It is necessary under this assumption that the probability density $\rho(p_i, q_i, t)$, and hence also the phase density $D(p_i, q_i, t)$ [they are proportional to each other by Eqs.(4.8)and (4.5)], must contain *no explicit dependence on the time*

$$\frac{\partial D}{\partial t} = 0 \qquad (7.18)$$

Equations (6.10) and (6.32) reflect the condition (6.8) that the *total* time derivative of D must also vanish

$$\frac{dD}{dt} = \frac{\partial D}{\partial t} - [H, D]_{P.B.} = 0 \qquad (7.19)$$

A combination of these results leads to the condition

$$[H, D]_{P.B.} = 0 \qquad (7.20)$$

In *summary*, for stationary mean values for arbitrary functions $Q(p_i, q_i)$ of the dynamical variables, by which we choose to characterize thermodynamic equilibrium, the phase density D must contain no explicit dependence on the time and must have a vanishing Poisson Bracket with the hamiltonian; it is evident from Eq.(7.19) [and from the analysis of Eqs.(6.18) - (6.20)] that $D(p_i, q_i)$ must be a *constant of the motion.*

8. Constancy of the Distribution Function(*)

The condition of thermal equilibrium has led us to the requirement that the phase density $D(p_i, q_i)$ should have no explicit dependence on the time and should satisfy

$$\frac{dD}{dt} = -[H, D]_{P.B.} = 0 \qquad (8.1)$$

In short, $D(p_i, q_i)$ should be a constant of motion for the many-body system.

Since the hamiltonian $H(p_i, q_i)$ [see, for example, Eq.(1.13)] is such a constant, an obvious possible choice would be to choose $D = H$, or more generally

$$D = f(H) \tag{8.2}$$

where f is an arbitrary function. Indeed, one has then with $f' = df/dH$

$$\frac{\partial D}{\partial p_i} = f' \frac{\partial H}{\partial p_i}$$
$$\frac{\partial D}{\partial q_i} = f' \frac{\partial H}{\partial q_i} \tag{8.3}$$

and hence from Eqs.(6.19)-(6.20)

$$\frac{dD}{dt} = -[H, D]_{P.B.} = -f'[H, H]_{P.B.}$$
$$= 0 \tag{8.4}$$

In fact, the choice in Eq.(8.2) will be seen to lead to the relation between statistical mechanics and the laws of thermodynamics.

Depending on the nature of the system considered, there are, however, other possibilities. For example, in the case of a system not subjected to any external forces, the components P_i with $i = (x, y, z)$ of the total momentum are likewise constants of motion, so that D could be an arbitrary function of H as well as of P_x, P_y, and P_z. This is *not* the case if the system is held in a fixed container, since its momentum may change by interaction with the walls.

Another example of special interest for later purposes is that where the total system consists of two non-interacting subsystems A and B. If the momenta and coordinates of the two systems are denoted by (p_{jA}, q_{jA}) and (p_{kB}, q_{kB}) respectively, then the hamiltonion of the total system is

$$H = H_A(p_{jA}, q_{jA}) + H_B(p_{kB}, q_{kB}) \tag{8.5}$$

and H_A and H_B are each constants of motion. Besides their sum H, the difference

$$\Delta = H_A - H_B \tag{8.6}$$

is therefore another constant of motion, so that one may choose

$$D = D(H, \Delta) \tag{8.7}$$

as an arbitrary function of both. Even the slightest interaction between A and B is sufficient, however, to deny this possibility, since it permits an energy transfer between the two systems; although it takes a longer time for Δ to change appreciably the smaller the interaction, Δ is then no longer a constant of motion.

To envisage the full scope of possibilities, go back to Eq.(6.6) with the requirement of Eq.(7.18) taken into account through the absence of an explicit time dependence. The corresponding equality then reduces to the relation

$$D[x(t)] = D(x_0) \tag{8.8}$$

where t appears in the role of a parameter for the sequence of dynamical variables along the phase orbit through the point x_0. This relation indicates that *the general solution of Eq.(8.1) is obtained by assigning to each phase orbit its own fixed value of D, which then remains constant along that phase orbit.*

For the claim of a solution having necessarily the form of Eq.(8.2), one is thus forced to postulate that, for a given value of H, there cannot be more than a single phase orbit, or expressed in geometrical terms, that the phase orbit passes through all the points on its energy surface. A system that would conform to such a possibility is called "*ergodic,*" and the acceptance of Eq.(8.2) as a consequence of equilibrium can in principle be based upon the hypothesis that the macroscopic systems under consideration are, in fact, ergodic.

This hypothesis, however, not only denies the existence of separate phase orbits on the same energy surface, distinguished from each other by different values of the constants of the motion other than the hamiltonian, but it is highly unrealistic under any circumstances to expect that a one-dimensional line completely fills a surface of $2n - 1$ dimension for $n > 1$. The fundamental objection is avoided if, instead of being ergodic, a system is merely assumed to be "*quasi-ergodic*" in the sense that the phase orbit comes arbitrarily close to any point on the energy surface. Although the distinction is significant from a formal point of view, neither the ergodic nor the rigorous quasi-ergodic hypothesis bears upon the physical situation faced in the actual observation of macroscopic systems. Even the case of strict quasi-ergodicity could be verified only by ascertaining the motion of the system with unlimited accuracy and over an infinitely long time.†

Let us examine the ergodic hypothesis in more detail. To rigorously satisfy Eq.(8.2) $D(p_i, q_i)$ must have the same value for all points (p_i, q_i)

---

† The second Bloch manuscript ends here. Much of the remainder of Sec. 8, and all of Sec. 9, are taken from the first draft.

of the phase space for which the energy $H(p_i, q_i)$ has a given value E. The equation

$$H(p_i, q_i) = E \qquad (8.9)$$

can be geometrically interpreted to define a $(2n-1)$-dimensional "surface" in the $2n$-dimensional phase-space. Choosing any point on this "energy surface" at a given time and following its motion, the corresponding phase orbit is then confined to lie on this energy surface. Considering further that D, whatever its form, shall be a constant of motion, it is thus required to have the same value on all points of the energy surface through which the phase orbit passes.

Assuming, then, that in the course of time the variables (p_i, q_i) will take every set of values compatible with Eq.(8.9), so that the phase orbit will pass through *all* points on the energy surface, D would indeed have the same value on the whole energy surface. D could then be arbitrarily chosen for any energy surface defined by a given value of E in Eq.(8.9). In that case D could depend only on E, thus justifying the equation

$$D = f(E) \qquad (8.10)$$

and hence Eq.(8.2), where the function f is still arbitrary.

There is no difficulty in maintaining the ergodic hypothesis for a system with a single degree of freedom, that is, for $n = 1$, since in that case $2n - 1 = 1$ and the "energy surface" degenerates into a one-dimensional line. For example, in the case of the harmonic oscillator with

$$H = \frac{p^2}{2m} + \frac{1}{2}m\omega^2 q^2 = E \qquad (8.11)$$

the energy surface has the form of an ellipse [Figure (8.1)]; this ellipse is covered by the motion of a system during each period of the oscillator and the energy surface is thus *identical* with the phase orbit in this case.

For $n = 2$ and $2n - 1 = 3$, however, the "energy surface" would already be three-dimensional, whereas the phase orbit still remains a one-dimensional line [Figure (8.2)]. It is highly unrealistic to assume that such a line should completely fill a three-dimensional continuum. Furthermore, even if this were possible, it would require the phase orbit to be infinitely long to very high order. To verify whether or not a system satisfies the assumption, one would therefore have to follow its motion during an infinitely long time, which, in practice, is evidently meaningless. This situation is even more extreme for the very large number n of degrees of freedom encountered in macroscopic systems.

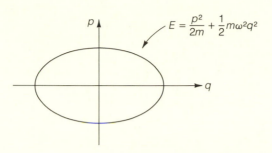

Figure 8.1 The "energy surface" for a one-dimensional simple harmonic oscillator. In this case the energy surface is identical to the phase orbit.

It should be noted that the ergodic hypothesis, although perhaps unrealistic, is not inconsistent with the general principles of mechanics in the following sense. A phase orbit is uniquely specified by giving the initial values (x_0) at the time t_0 as illustrated in Figure 8.2 and Eq. (8.8). There are as many phase orbits as sets of initial values (x_0), and phase orbits cannot cross. The subsequent orbit is then determined by integrating Hamilton's equations, starting from (x_0) at t_0. Now at a later time t, the phase orbit will be at a new point $[x(t)]$ that may be taken to define a new initial value (x_0'). At time t_0, one would say that this point (x_0') defines a new and distinct phase orbit, but now this orbit is actually traced as a continuation of the original phase orbit, albeit at a later (and conceivably much later) time.

There are physical systems that clearly do *not* exhibit ergodic behavior. Imagine a collection of non-interacting particles, no matter how many, moving with the same velocity perpendicular to two walls of a cubical box, with which they undergo elastic collisions. The particles will continue to bounce back and forth, and the phase orbit cannot pass through all points on the energy surface in phase space.

It is possible, nevertheless, to effectively justify Eq.(8.2) by relaxing the assumption of ergodicity and demanding that the system be merely *"quasi-ergodic"* in a certain sense. Without attempting a rigorous definition of this property the underlying ideas can be qualitatively explained. Consider a given phase orbit on an energy surface and an arbitrary point P' on the same suface, but not necessarily on the same phase orbit [Figure (8.3)]. It is then sufficient to assume that somewhere

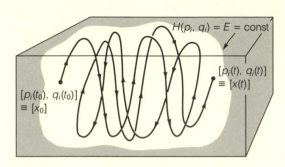

Figure 8.2   Schematic representation of a phase orbit lying on the "energy surface" in the case where the dimension of phase space is $2n \geq 4$ and of the energy surface is $2n - 1 \geq 3$.

along a phase orbit there is a point P, "close" to P' in the sense that with $P(p_i, q_i)$ and $P'(p_i + \Delta p_i, q_i + \Delta q_i)$ one has

$$
\begin{aligned}
|\Delta p_i| &\leq \varepsilon_i \\
|\Delta q_i| &\leq \delta_i
\end{aligned}
$$

(8.12)

so that P' is closer to P the smaller $(\varepsilon_i, \delta_i)$. (The ergodic hypothesis would require $\varepsilon_i = 0, \delta_i = 0$.)

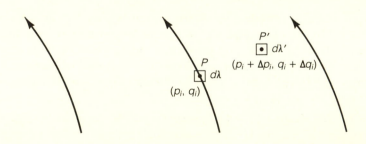

Figure 8.3   Quantities used in defining "sufficiently quasi-ergodic" behavior.

Let ρ and ρ' denote the probability density at P and P' respectively so that $\rho d\lambda$ and $\rho' d\lambda'$ represent the probabilities of finding the system within the volume elements $d\lambda$ and $d\lambda'$ around P and P'. To assume a notable difference between these probabilities would be meaningful in practice only if the observed properties of the system at P and P' were significantly different. This would clearly not be the case, for example, if due to the experimental error, p_i and q_i could not be measured with an accuracy comparable to ε_i and δ_i respectively. Depending upon the questions asked, the distinction between the properties at P and P' may be irrelevant for even larger differences Δp_i and Δq_i and hence even larger values of ε_i and δ_i. The terminology used to distinguish between finite and arbitrarily small values is to speak about *"coarse grain"* and *"fine grain"* descriptions of a system.

One is thus led to expect equal probabilities, and thus $\rho = \rho'$, if ε_i and δ_i are only sufficiently small to render the difference in properties at P and P' insignificant. In this sense, the system may be called "sufficiently quasi-ergodic," whereby it depends upon the circumstances, *how* small ε_i and δ_i have to be chosen or *how* "coarse-grain" a description of the system is sufficient for the purposes. In particular, for obtaining the average value of a quantity $Q(p_i, q_i)$ it may be sufficient to choose ε_i and δ_i small enough so that Q changes little upon a change satisfying Eqs.(8.12). Considering further that D and hence $\rho = D/\nu$ has the same value at all points P on the phase orbit, and that to any other point P' one can associate a point P such that $\rho' = \rho$, it thus follows that ρ and hence D can indeed be assumed to have the same value everywhere on the energy surface, so that Eq.(8.2) is in fact effectively justified.

Let us attempt a *summary*. One can say that for the thermodynamic quantities of experimental interest, "sufficiently quasi-ergodic" behavior and "coarse-grain averaging" permit the integral

$$\bar{Q} \;=\; \int Q(p_i, q_i)\rho(p_i, q_i)d\lambda \tag{8.13}$$

to be replaced by a sum over the set of points $\{P\}$

$$\bar{Q} \;=\; \sum_{\{P\}} Q(p_i, q_i)\rho(p_i, q_i) \prod_j \Delta p_j \Delta q_i \tag{8.14}$$

where the sum goes first over points that lie on a finite segment of an arbitrarily chosen phase orbit on a given "energy surface," and then over a finite set of distinct "energy surfaces" (Figure 8.4). The validity of this replacement can, of course, be checked in specific examples. Equation

(8.2) is evidently satisfied if this replacement is justified, and we shall henceforth assume this to be the case.

It remains to determine the *form* of the function f in Eq.(8.2), and we proceed to do that in the next section.

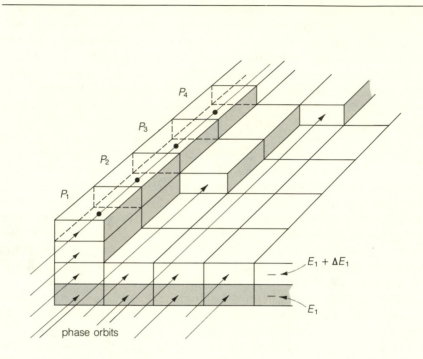

Figure 8.4 Schematic illustration of "sufficiently ergodic coarse-grain" averaging in phase space.

9. The Canonical Distribution (*)

As long as one considers only the equilibrium of an isolated system, it is sufficient to postulate Eq.(8.2) with an arbitrary dependence of the phase density D upon the hamiltonian H. The same holds for the probability density

$$\rho \;=\; \rho(H) \tag{9.1}$$

as an arbitrary function of H.

In order to determine the form of this function, it is necessary, however, to invoke a fact of thermodynamic equilibrium that concerns the mutual equilibrium of several systems and leads to the introduction of the concept of *temperature*. Briefly stated, it says that all parts of a system in thermodynamic equilibrium have the same temperature or, more specifically, that the equilibrium of two systems in contact with each other is reached at a temperature common to both. In fact, temperature is rigorously defined as being that which two systems in contact and in thermodynamic equilibrium have in common. Indeed, a thermometer consists merely of a system, brought into contact with another system, with one of its properties, such as the height of a liquid column in a capillary reached in equilibrium, used as a measure of the temperature. The scale of temperature is thereby left entirely open, and the calibration of two different thermometers against each other is achieved merely by stating that they shall read the same temperature when they have reached equilibrium in contact with each other, or with a common third system. If sufficiently large, the latter can be used as a "heat reservoir" that exhibits a negligible temperature change upon equilibrium reached in contact with another relatively small system.

It is necessary, therefore, to consider again, as in Sec. 8, two systems A and B with hamiltonians $H_A(p_{jA}, q_{jA})$ and $H_B(p_{kB}, q_{kB})$. For each separately in equilibrium, the probability to find A within $d\lambda_A = \prod_j dp_{jA} dq_{jA}$ and B within $d\lambda_B = \prod_k dp_{kB} dq_{kB}$ is then

$$dP_A = \rho_A(H_A) d\lambda_A \qquad (9.2)$$

and

$$dP_B = \rho_B(H_B) d\lambda_B \qquad (9.3)$$

respectively, where ρ_A and ρ_B can be independently chosen as arbitrary functions. Now consider A and B combined as a single system, denoted by C. The probability dP_C of finding *both* A within $d\lambda_A$ *and* B within $d\lambda_B$ and, hence, to find C within $d\lambda_C = \prod_j dp_{jA} dq_{jA} \prod_k dp_{kB} dq_{kB}$, so that

$$d\lambda_C = d\lambda_A d\lambda_B \qquad (9.4)$$

is then given by the product

$$dP_C = dP_A dP_B \qquad (9.5)$$

or from Eqs.(9.2) - (9.4)

$$dP_C = \rho_A(H_A)\rho_B(H_B)d\lambda_C \qquad (9.6)$$

As long as A and B are uncoupled, the hamiltonian of the combined system C is

$$H_C = H_A + H_B \qquad (9.7)$$

A "contact" between A and B can be described, in mechanical terms, by introducing a potential energy

$$h_{A,B} = h(q_{jA}, q_{kB}) \qquad (9.8)$$

chosen to permit a transfer of energy between A and B, so that now

$$H_C = H_A + H_B + h_{A,B} \qquad (9.9)$$

The fact that this contact shall not significantly alter the properties of the combined system further requires that

$$|h_{A,B}| << |H_A + H_B| \qquad (9.10)$$

Although an appreciable transfer of energy between A and B will take longer the smaller $h_{A,B}$, it is possible to assume that thermodynamic equilibrium can be established in the sense that not only A and B separately, but also C, are in equilibrium. In that case, besides Eqs.(9.2) and (9.3), one has then

$$dP_C = \rho_C(H_C)d\lambda_C \qquad (9.11)$$

for the probability to find C with $d\lambda_C$. Although H_C is strictly given by Eq.(9.9), it is allowed, in view of Eq.(9.10), to use Eq.(9.7) instead; it follows from Eq. (9.6) that

$$\rho_C(H_A + H_B) = \rho_A(H_A)\rho_B(H_B) \qquad (9.12)$$

or, for a given energy $E_A = H_A$ and $E_B = H_B$ of systems A and B respectively,

$$\rho_C(E_A + E_B) = \rho_A(E_A)\rho_B(E_B) \qquad (9.13)$$

Whereas the function ρ_A and ρ_B can so far be chosen arbitarily, their choice thus determines that of the function ρ_C. Since Eq.(9.13) holds for arbitrary values of E_A and E_B, it also remains valid for a differential change dE_A and dE_B of either, so that, upon forming the partial derivatives on both sides of Eq.(9.13),

$$\rho_C'(E_A + E_B) = \rho_A'(E_A)\rho_B(E_B) \tag{9.14}$$

and

$$\rho_C'(E_A + E_B) = \rho_A(E_A)\rho_B'(E_B) \tag{9.15}$$

Where the prime indicates the derivatives of $\rho_{A,B,C}$ with respect to their single arguments. Thus

$$\rho_A'(E_A)\rho_B(E_B) = \rho_A(E_A)\rho_B'(E_B) \tag{9.16}$$

or

$$\frac{\rho_A'(E_A)}{\rho_A(E_A)} = \frac{\rho_B'(E_B)}{\rho_B(E_B)} \tag{9.17}$$

If the common value of ρ_A'/ρ_A and ρ_B'/ρ_B is denoted by r, one has therefore

$$\frac{\rho_A'(E_A)}{\rho_A(E_A)} = r \tag{9.18}$$

and

$$\frac{\rho_B'(E_B)}{\rho_B(E_B)} = r \tag{9.19}$$

It follows from Eq.(9.18) that $\partial r/\partial E_B = 0$ and from Eq.(9.19) that $\partial r/\partial E_A = 0$, so r depends neither on E_A nor on E_B. Furthermore, according to Eq.(9.18), the value of r does not depend on the choice of the system B and, according to Eq.(9.19), it does not depend on the choice of the system A either. The r is therefore a *common constant* for all systems in statistical equilibrium with the property that it remains unchanged upon contact between any two of them. On the other hand, thermodynamic equilibrium implies that the only quantity that such systems have in common is their *temperature*, and hence r may be expected to depend only upon the temperature. In particular, one of the systems may be considered as a heat reservoir in equilibrium, with a given value of r characteristic of its temperature, so that the equilibrium of any system in contact with this heat reservoir is characterized by the same value of r. It is customary to write

$$r \equiv -\beta \tag{9.20}$$

Irrespective of the nature of A and B, one has then from Eqs.(9.18) or (9.19)

$$\frac{\rho'(E)}{\rho(E)} = -\beta \qquad (9.21)$$

and hence

$$\rho(E) = Ce^{-\beta E} \qquad (9.22)$$

Since the energy E generally has a lower but no upper limit, one has to have $\beta > 0$ in order to prevent ρ from indefinitely increasing with increasing E. Since E represents the set of values (p_i, q_i) for which $H(p_i, q_i) = E$, one has from Eq.(9.22) for the dependence of ρ on the variables (p_i, q_i)

$$\rho(p_i q_i) = Ce^{-\beta H(p_i, q_i)} \qquad (9.23)$$

Since, in view of Eq.(4.9),

$$\int \rho(p_i, q_i) d\lambda = 1 \qquad (9.24)$$

the constant C is determined by

$$C \int e^{-\beta H(p_i, q_i)} d\lambda = 1 \qquad (9.25)$$

or

$$C = 1 / \int e^{-\beta H(p_i, q_i)} d\lambda \qquad (9.26)$$

where the integral extends over the whole phase space of the variables (p_i, q_i). Therefore

$$\rho(p_i, q_i) = \frac{e^{-\beta H(p_i, q_i)}}{\int e^{-\beta H(p_i, q_i)} d\lambda} \qquad (9.27)$$

Instead of the probability density, one may also consider the phase density D of an ensemble with ν members. Since from Eqs.(4.8) and (4.5)

$$D = \nu\rho \qquad (9.28)$$

one has

$$D(p_i, q_i) = D_0 e^{-\beta H(p_i, q_i)} \qquad (9.29)$$

with

$$D_0 = \frac{\nu}{\int e^{-\beta H(p_i, q_i)} d\lambda} \tag{9.30}$$

Equations (9.27) or (9.29, 9.30) are basic for the statistical description of thermodynamic equilibrium. In the terminology of Gibbs, the corresponding *ensemble* is called *"canonical"* and $\theta \equiv 1/\beta$ is called the *"modulus"* of the canonical ensemble.

10. Thermodynamic Functions

In the formulation of thermodynamics it is important to recognize the existence of certain quantities with the property that, going over from one equilibrium state to another, their change is only determined by that of the thermodynamic variables, irrespective of the intermediate states through which one had to pass. These quantities can then be considered as functions of the thermodynamic variables, called *thermodynamic functions* (or "state functions").

The first law of thermodynamics, which implies conservation of energy, recognizes the energy as such a quantity. From the point of view of mechanics, this is already assured by Hamilton's equations, but *statistically*, one can only make statements of probability about the actual value of the energy. For large systems, however, it can be expected to be found close to the *average value*, which can thus be regarded as "the" energy E. Thus

$$E \equiv \bar{H} = \int H \rho d\lambda$$
$$E = \frac{\int H e^{-\beta H} d\lambda}{\int e^{-\beta H} d\lambda} \tag{10.1}$$

Since it will be important to consider work done upon the system, H must be allowed to depend on a set of external parameters a_j. Thus

$$H = H(p_i, q_i, a_j) \tag{10.2}$$

Then

$$E = E(\beta, a_j) \tag{10.3}$$

Since

$$H e^{-\beta H} = -\frac{\partial}{\partial \beta} e^{-\beta H} \tag{10.4}$$

one can write

$$E = \frac{-\int \frac{\partial}{\partial \beta} e^{-\beta H} d\lambda}{\int e^{-\beta H} d\lambda} = \frac{\frac{-\partial}{\partial \beta} \int e^{-\beta H} d\lambda}{\int e^{-\beta H} d\lambda} \tag{10.5}$$

This expression can be written more compactly in terms of the *"partition function"* defined by

$$\mathcal{Z} \equiv \int e^{-\beta H} d\lambda = \mathcal{Z}(\beta, a_j) \tag{10.6}$$

Thus

$$E = \frac{\frac{-\partial \mathcal{Z}}{\partial \beta}}{\mathcal{Z}} = -\frac{\partial ln \mathcal{Z}}{\partial \beta} \tag{10.7}$$

We note that both the energy and partition function are functions of (β, a_j).

The *first law of thermodynamics* states that

$$\Delta E = \Delta W + \Delta Q \tag{10.8}$$

where ΔW and ΔQ *will* depend on the intermediate states, whereas ΔE does not. For differentially small changes Eq.(10.8) can be written as

$$dE = \delta W + \delta Q \tag{10.9}$$

where δW and δQ are *not* differentials of thermodynamic functions W or Q, but dE is. The differential of Eq.(10.3) in fact gives

$$dE = \frac{\partial E}{\partial \beta} d\beta + \sum_j \frac{\partial E}{\partial a_j} da_j \tag{10.10}$$

The insertion of Eq.(10.7) then yields

$$dE = -\left(\frac{\partial^2 ln \mathcal{Z}}{\partial \beta^2} d\beta + \sum_j \frac{\partial^2 ln \mathcal{Z}}{\partial \beta \, \partial a_j} da_j \right) \tag{10.11}$$

Now we do not yet know what is meant by δQ, but we do know from our previous discussion that for a given H the work element δW can be written in terms of the change in external parameters according to Eq.(1.64)

$$\delta W = \sum_j \frac{\partial H}{\partial a_j}(p_i, q_i, a_j) da_j \tag{10.12}$$

Since we do not know the values of the (p_i, q_i) in this expression, we can only make *probability* statements about the work done on the system, that is, one can only guess at δW. Again, however, for macroscopic systems δW is expected to be very close to the mean value. Therefore we define the work as the statistical average and identify this as the quantity that we would actually measure

$$\delta W \equiv \sum_j \overline{\left(\frac{\partial H}{\partial a_j}\right)} da_j = \sum_j \left(\int \frac{\partial H}{\partial a_j} \rho \, d\lambda\right) da_j$$

$$= \sum_j \left(\frac{\int \frac{\partial H}{\partial a_j} e^{-\beta H} d\lambda}{\int e^{-\beta H} d\lambda}\right) da_j \tag{10.13}$$

The use of the identity

$$\frac{\partial H}{\partial a_j} e^{-\beta H} = -\frac{1}{\beta} \frac{\partial}{\partial a_j} e^{-\beta H} \tag{10.14}$$

allows Eq.(10.13) to be rewritten as

$$\delta W = -\sum_j \frac{1}{\beta} \left(\frac{\frac{\partial}{\partial a_j} \int e^{-\beta H} d\lambda}{\int e^{-\beta H} d\lambda}\right) = -\sum_j \frac{1}{\beta} \frac{\frac{\partial Z}{\partial a_j}}{Z} da_j \tag{10.15}$$

So, finally,

$$\delta W = -\frac{1}{\beta} \sum_j \frac{\partial \ln Z}{\partial a_j} da_j \tag{10.16}$$

This expression provides a definition of the infinitesimal, macroscopic work done on the system.

We are now in a position to determine *statistically* what is meant by δQ, for

$$\delta Q = dE - \delta W$$

$$= \left(\frac{\partial E}{\partial \beta} d\beta + \sum_j \frac{\partial E}{\partial a_j} da_j\right) - \delta W \tag{10.17}$$

The use of Eqs.(10.11) and (10.16) allows this result to be written in terms of the partition function

$$\delta Q = -\frac{\partial^2 \ln Z}{\partial \beta^2} d\beta - \sum_j \left(\frac{\partial^2 \ln Z}{\partial \beta \, \partial a_j} da_j - \frac{1}{\beta} \frac{\partial \ln Z}{\partial a_j} da_j\right) \tag{10.18}$$

Note that we have taken the equilibrium expression of δW for any values of a_j. In changing these values, we assume therefore that we deal only with *reversible* changes.

Let us combine these results in a slightly different fashion. The total differential of the logarithm of the partition function is Eq.(10.6) is given by

$$dln\mathcal{Z} = \frac{\partial ln\mathcal{Z}}{\partial \beta}d\beta + \sum_j \frac{\partial ln\mathcal{Z}}{\partial a_j}da_j \qquad (10.19)$$

The substitution of the previous results for the energy and infinitesimal work elements in Eqs.(10.7) and (10.16) then yields

$$
\begin{aligned}
dln\mathcal{Z} &= -Ed\beta - \beta\delta W \\
&= -\beta\delta W - d(E\beta) + \beta dE
\end{aligned} \qquad (10.20)
$$

This may be rewritten as

$$
\begin{aligned}
d(ln\mathcal{Z} + \beta E) &= \beta(dE - \delta W) \\
d(ln\mathcal{Z} + \beta E) &= \beta\delta Q
\end{aligned} \qquad (10.21)
$$

Since the lefthand side of this expression is a total differential, so must be the righthand side. Hence we have found that the combination $\beta\delta Q$ is a thermodynamic, or state function. We can interpret this particular thermodynamic function with the aid of the second law of thermodynamics.

Now the *second law of thermodynamics* refers to the gain of heat δQ in a reversible process by stating that

$$dS = \frac{\delta Q}{T} \qquad (10.22)$$

where T is defined to be the "absolute temperature," and the *"entropy"* S is a thermodynamic function. The second law now establishes a relation between T and β. Let

$$\frac{1}{T} \equiv g(\beta) \qquad (10.23)$$

Then

$$dS = g(\beta)\delta Q \qquad (10.24)$$

Since entropy is a thermodynamic function we have

$$S = S(\beta, a_j) \qquad (10.25)$$

and the total differential of this expression gives

$$dS = \frac{\partial S}{\partial \beta}d\beta + \sum_j \frac{\partial S}{\partial a_j}da_j \qquad (10.26)$$

Comparison with Eqs.(10.24) and (10.18) then yields the expressions

$$\frac{\partial S}{\partial \beta} = -g(\beta)\frac{\partial^2 ln Z}{\partial \beta^2} \tag{10.27}$$

$$\frac{\partial S}{\partial a_j} = -g(\beta)\left(\frac{\partial^2 ln Z}{\partial \beta\, \partial a_j} - \frac{1}{\beta}\frac{\partial ln Z}{\partial a_j}\right) \tag{10.28}$$

where the other variables in Eq.(10.25) are to be held fixed in taking the partial derivatives. Thus Eq.(10.28) can be rewritten as

$$\frac{\partial S}{\partial a_j} = -\frac{\partial}{\partial a_j}g(\beta)\left[\frac{\partial ln Z}{\partial \beta} - \frac{1}{\beta}ln Z\right] \tag{10.29}$$

For a state function

$$S = S(\beta, a_j) \tag{10.30}$$

it must be true that

$$\frac{\partial^2 S}{\partial a_j \partial a_i} = \frac{\partial^2 S}{\partial a_i \partial a_j} \tag{10.31}$$

$$\frac{\partial^2 S}{\partial a_j \partial \beta} = \frac{\partial^2 S}{\partial \beta \partial a_j} \tag{10.32}$$

The insertion of Eqs.(10.27) and (10.28) in this last relation gives

$$g(\beta)\frac{\partial^3 ln Z}{\partial \beta^2 \partial a_j} = \frac{\partial}{\partial \beta}g(\beta)\left[\frac{\partial^2 ln Z}{\partial \beta\, \partial a_j} - \frac{1}{\beta}\frac{\partial ln Z}{\partial a_j}\right] \tag{10.33}$$

and explicit evaluation of the derivative on the righthand side yields

$$\begin{aligned} g(\beta)\frac{\partial^3 ln Z}{\partial \beta^2 \partial a_j} &= g(\beta)\left[\frac{\partial^3 ln Z}{\partial \beta^2 \partial a_j} - \frac{1}{\beta}\frac{\partial^2 ln Z}{\partial \beta\, \partial a_j} + \frac{1}{\beta^2}\frac{\partial ln Z}{\partial a_j}\right] \\ &\quad + g'(\beta)\left[\frac{\partial^2 ln Z}{\partial \beta\, \partial a_j} - \frac{1}{\beta}\frac{\partial ln Z}{\partial a_j}\right] \end{aligned} \tag{10.34}$$

This implies

$$-g(\beta)\frac{\partial}{\partial \beta}\left(\frac{1}{\beta}\frac{\partial ln Z}{\partial a_j}\right) + g'(\beta)\left[\frac{\partial^2 ln Z}{\partial \beta\, \partial a_j} - \frac{1}{\beta}\frac{\partial ln Z}{\partial a_j}\right] = 0 \tag{10.35}$$

or

$$\frac{g'(\beta)}{g(\beta)} = \frac{\left[\frac{1}{\beta}\frac{\partial^2 ln Z}{\partial \beta\, \partial a_j} - \frac{1}{\beta^2}\frac{\partial ln Z}{\partial a_j}\right]}{\left[\frac{\partial^2 ln Z}{\partial \beta\, \partial a_j} - \frac{1}{\beta}\frac{\partial ln Z}{\partial a_j}\right]} = \frac{1}{\beta} \tag{10.36}$$

The unknown function $g(\beta)$ must thus satisfy the relation

$$\frac{g'(\beta)}{g(\beta)} = \frac{1}{\beta} \tag{10.37}$$

If $g(\beta)$ satisfies this relation, then dS will be the perfect differential of a thermodynamic function. The solution to Eq.(10.37) is

$$ln\, g = ln\, \beta + \text{const.} \tag{10.38}$$

or

$$g(\beta) = k_B\beta \tag{10.39}$$

where k_B is a constant. So from Eq.(10.23)

$$\beta = \frac{1}{k_BT} \tag{10.40}$$

Where T is the absolute temperature. Since the temperature scale is otherwise arbitrary, the constant k_B is arbitrary; if T is measured in degrees Kelvin, then k_B can be determined in the following way.

We will show later (Sec. 13) that an ideal gas satisfies the equation of state

$$pV = Nk_BT \tag{10.41}$$

where N is the number of molecules. Thus

$$\begin{aligned} pv &= Ak_BT \\ &= RT \end{aligned} \tag{10.42}$$

where v is the molar volume, $R \equiv Ak_B$ is the gas constant, and A is Avogadro's number. The measured values of these quantities

$$\begin{aligned} R &= 8.316 \times 10^7 \text{ erg}(^0K)^{-1}\text{mole}^{-1} \quad ; \text{ gas constant} \\ A &= 6.022 \times 10^{23} \text{ mole}^{-1} \qquad\qquad\; ; \text{ Avogadro's number} \end{aligned} \tag{10.43}$$

allow a determination of k_B, which is *Boltzmann's constant*.

$$k_B = 1.381 \times 10^{-16} \text{ erg}(^0K)^{-1} \qquad ; \text{ Boltzmann's constant} \tag{10.44}$$

Note that $\beta = 1/k_BT$ has dimensions of $(\text{energy})^{-1}$, so that the quantity βH appearing in the partition function in Eq.(10.6) is dimensionless.

We are now in a position to derive an expression for the entropy. The identification of the absolute temperature in Eq.(10.23) implies that Eqs.(10.27) and (10.29) take the form

$$\frac{\partial S}{\partial \beta} = -k_B\beta\frac{\partial^2 ln\mathcal{Z}}{\partial \beta^2} = k_B\frac{\partial}{\partial \beta}\left[ln\mathcal{Z} - \beta\frac{\partial ln\mathcal{Z}}{\partial \beta}\right] \tag{10.45}$$

$$\frac{\partial S}{\partial a_j} = -\frac{\partial}{\partial a_j} k_B \beta \left[\frac{\partial ln\mathcal{Z}}{\partial \beta} - \frac{ln\mathcal{Z}}{\beta} \right] = k_B \frac{\partial}{\partial a_j} \left[ln\mathcal{Z} - \beta \frac{\partial ln\mathcal{Z}}{\partial \beta} \right] \quad (10.46)$$

The partial derivatives are now those of a pure function, as in Eq.(10.25); the choice of $g(\beta)$ in Eq.(10.23) is unique, and it must be so in order to make S a thermodynamic function. The entropy can thus be identified as

$$S = k_B \left[ln\mathcal{Z} - \beta \frac{\partial ln\mathcal{Z}}{\partial \beta} \right] + S_0 \quad (10.47)$$

where S_0 is a constant independent of β and a_j. The thermodynamic definition of S relates only to *changes* in S so we can always add a constant S_0. Note that there is also an arbitrary constant in E defined in Eq.(10.1), since an additive constant in H does not change Hamilton's equations

$$\dot{p}_i = -\frac{\partial H}{\partial q_i} \quad ; \quad \dot{q}_i = \frac{\partial H}{\partial p_i} \quad (10.48)$$

Finally, we observe that the expression for the entropy in Eq.(10.47) is identical to that discussed previously in Eqs.(10.21) and (10.7).

11. The Partition Function

In summary, we have finally arrived at the main connection between statistical mechanics and thermodynamics. There is a quantity called *entropy* that can be derived from the *partition function*

$$\mathcal{Z} = \int e^{-H/k_B T} d\lambda \quad (11.1)$$

The partition function relates to the properties of the *microscopic* system. It provides all the thermodynamic parameters of a *macroscopic* sample.

Another important thermodynamic function is the *free energy* defined by †

$$F \equiv E - TS \quad (11.2)$$

It is evident from the definition that this is a state function. To understand the importance of the free energy, consider a reversible *isothermal* process, that is, one carried out at constant temperature. The first and second laws of thermodynamics state that

$$\begin{aligned} dF &= dE - TdS \\ dF &= \delta Q + \delta W - \delta Q \\ &= \delta W \quad\quad\quad ; T = \text{const.} \end{aligned} \quad (11.3)$$

---

† This function is usually called the Helmholtz free energy. The Gibbs' free energy is defined by G = E + PV−TS.

Thus in a reversible isothermal process, the change in the state function F is identical to the work done on the system. Note that in this particular case, the work increment becomes a perfect differential.

Substitution of Eqs.(10.7) and (10.47) allows the free energy to be written in terms of the partition function

$$F = -\frac{\partial ln \mathcal{Z}}{\partial \beta} - \frac{1}{\beta}\left[ln \mathcal{Z} - \beta\frac{\partial ln \mathcal{Z}}{\partial \beta} \right] - \frac{S_0}{k_B \beta} \qquad (11.4)$$

This expression reduces to

$$F = -\frac{1}{\beta}ln\mathcal{Z} - \frac{S_0}{k_B\beta}$$
$$F = -\frac{1}{\beta}\left(ln\mathcal{Z} + \frac{S_0}{k_B}\right) \qquad (11.5)$$

The free energy is thus simply expressed in terms of the partition function. Exponentiation of this relation gives

$$\mathcal{Z} = \mathcal{Z}_0 e^{-\beta F} \qquad (11.6)$$

where the overall constant factor in the partition function is defined by

$$\mathcal{Z}_o \equiv e^{-S_0/k_B} \qquad (11.7)$$

Thus, conversely, the partition function is very simply expressed in terms of the free energy.

12. The Statistical Significance of Entropy

As we have seen, the thermodynamic concept of *energy* has a mechanical significance. In contrast, the concept of *entropy* is really statistical in nature. In this section we pursue that connection.

Knowledge of the probability density

$$\rho = \frac{e^{-\beta H}}{\int e^{-\beta H} d\lambda} \qquad (12.1)$$

gives the "betting odds" for finding a system at a particular point in phase space in the canonical ensemble; however, one is still left with a considerable degree of "ignorance" about what to expect for a given system. How can we measure this ignorance? One way would be to specify how big a volume in phase space one must take in order to have some high probability, say greater than a 99.99% chance, of the system being in that volume . The smaller the volume required, the closer one could then specify the actual values of the dynamical variables of the system. This volume is called the *"phase extension"* by Gibbs.

In order to give this concept a more rigorous definition, let σ_1 and σ_2 denote two energy surfaces in phase space defined by

$$H(p_i, q_i) \;=\; E_{1,2} \tag{12.2}$$

as illustrated in Figure 12.1.

Consider two intermediate surfaces defined by E' and $E' + dE'$ and let the phase space volume between these surfaces be defined by

$$d\Omega \;\equiv\; \phi(E')dE' \tag{12.3}$$

The volume between the surfaces σ_1 and σ_2 is then given by

$$\Omega \;=\; \int_{E_1}^{E_2} \phi(E')dE' \tag{12.4}$$

The probability $P(E_1, E_2)$ that the system will be found within this volume is given by Eqs.(12.1) and (12.3) as

$$P(E_1, E_2) \equiv \frac{\int_{\sigma_1}^{\sigma_2} e^{-\beta H} d\lambda}{\int e^{-\beta H} d\lambda} = \frac{\int_{E_1}^{E_2} \phi(E')e^{-\beta E'} dE'}{\int_{E_0}^{\infty} \phi(E')e^{-\beta E'} dE'} \tag{12.5}$$

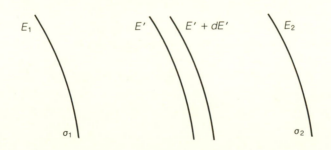

Figure 12.1 Energy surfaces in phase space used in the definition of the "phase extension."

where E_0 denotes the lowest possible energy of the system. The denominator in this expression is just the partition function

$$\mathcal{Z} = \int e^{-\beta H} d\lambda = \int_{E_0}^{\infty} \phi(E')e^{-\beta E'} dE' \qquad (12.6)$$

Let us get an idea of the structure of these quantities by looking at the integrand $\phi(E') \exp{(-\beta E')}$ sketched in Figure 12.2.

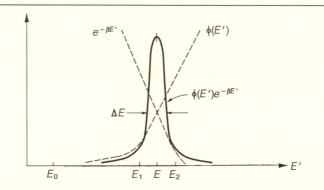

Figure 12.2 Phase extension and probability distribution vs. energy.

Now it is evident from Eq.(10.1) that the thermodynamic energy is to be identified with the mean value taken over the canonical ensemble

$$E = \frac{\int E'\phi(E')e^{-\beta E'} dE'}{\int \phi(E')e^{-\beta E'} dE'} \qquad (12.7)$$

As discussed in Sec. 5, to be an appropriate thermodynamic variable, the integrand in this expression must be sharply peaked. Thus we expect $\phi(E')$ to be a strongly increasing function for a macroscopic system; the product $\phi(E') \exp(-\beta E')$ is then damped at large E' by the decreasing exponential. For thermodynamics, the peak in Figure 12.2 must be a narrow one. In fact, in Sec. 5 it was shown that for a macroscopic system composed of a large number N of subsystems, the region of phase space containing essentially unit probability can be expected to satisfy the conditions†

$$E \sim N \qquad (12.8)$$

† The symbol $\sim$ means "is proportional to," and $\cong$ means "is approximately equal to."

$$\Delta E \sim \sqrt{N} \qquad (12.9)$$

Hence the peak will indeed be narrow

$$\left| \frac{\Delta E}{E} \right| << 1 \qquad (12.10)$$

Under these conditions, the thermodynamic energy of the system is sensibly given by the position of the maximum of the peak, which is located by the condition

$$\frac{\partial}{\partial E'} \left[\phi(E') e^{-\beta E'} \right] \Big|_{E'=E} = \frac{\partial \phi}{\partial E} e^{-\beta E} - \beta \phi(E) e^{-\beta E} = 0 \qquad (12.11)$$

This yields the relation

$$\frac{\partial ln\phi(E)}{\partial E} = \beta \qquad (12.12)$$

which determines the thermodynamic, or mean energy $E = E(\beta)$.

In order that the actual energy of the system lie in the included volume, the energies E_1 and E_2 should be chosen as indicated in Figure 12.2. One must choose

$$E - E_0 >> |E - E_1|, |E - E_2| > \Delta E \qquad (12.13)$$

If these conditions are satisfied, we can expect that the probability $P(E_1, E_2)$ will be very close to unity.

Now if the peak in Figure 12.2 is indeed very narrow, then the integral in the numerator of Eq.(12.5) can be written approximately as

$$\int_{E_1}^{E_2} \phi(E') e^{-\beta E'} dE' \cong \phi(E) e^{-\beta E} \Delta E \qquad (12.14)$$

where ΔE is an energy increment that measures the width of the distribution (Figure 12.2). In addition, if the peak in Figure 12.2 is very sharp, then the integral appearing in the *partition function* in Eq.(12.6) can similarly be written approximately as

$$\mathcal{Z} \cong \phi(E) e^{-\beta E} \Delta E \qquad (12.15)$$

The use of Eqs.(12.14) and (12.15) in Eq.(12.5) now yields $P(E_1, E_2) \cong 1$ provided the conditions of Eq.(12.13) are satisfied. Furthermore, if these conditions are satisfied, the *phase extension* itself in Eq.(12.4) can *also* be written approximately as (Figure 12.2)

$$\Omega \cong \phi(E) \Delta E \qquad (12.16)$$

The logarithm of the partition function is now immediately obtained from Eq.(12.15)

$$ln\mathcal{Z} \cong ln\phi(E) - \beta E + ln\Delta E \tag{12.17}$$

The entropy can then be evaluated from Eq.(10.47). One needs

$$\frac{\partial ln\mathcal{Z}}{\partial \beta} \cong \left(\frac{\partial ln\phi(E)}{\partial E} - \beta\right)\frac{\partial E}{\partial \beta} - E + \frac{\partial ln\Delta E}{\partial \beta} \tag{12.18}$$

The expression in parentheses vanishes by Eq.(12.12), and thus

$$S - S_0 = k_B\left(ln\mathcal{Z} - \beta\frac{\partial ln\mathcal{Z}}{\partial \beta}\right)$$
$$\cong k_B\left[ln\phi(E) + ln\Delta E - \beta\frac{\partial ln\Delta E}{\partial \beta}\right] \tag{12.19}$$

To proceed, let us return to the arguments in Sec. 5 and Eqs.(12.8-9) and examine them in more detail. For N loosely coupled subsystems, where the interaction energy is negligible, one can write

$$H \cong \sum_{s=1}^{N} h_s \tag{12.20}$$

The volume element in phase space always takes the form

$$d\lambda = \prod_{s=1}^{N} d\lambda_s \tag{12.21}$$

and hence the *partition function* for N loosely coupled subsystems *factors*

$$\mathcal{Z} = \int e^{-\beta H}d\lambda = \int \prod_{s=1}^{N} e^{-\beta h_s}d\lambda_s = \prod_{s=1}^{N} \int e^{-\beta h_s}d\lambda_s$$
$$\equiv \prod_{s=1}^{N} z_s \tag{12.22}$$

If the subsystems are identical, the factors $z_s = z$ will all be the same, and one finds

$$\mathcal{Z} = z^N \tag{12.23}$$

The logarithm of this expression then yields

$$ln\mathcal{Z} = Nlnz \tag{12.24}$$

The total energy of the system is given by Eq.(10.7)

$$E = -\frac{\partial ln\mathcal{Z}}{\partial\beta} = N\left(-\frac{\partial lnz}{\partial\beta}\right) \qquad (12.25)$$

The energy is thus an extensive quantity; it is proportional to the number of subsystems N as claimed in Eq.(12.8).

Consider next the mean square deviation of the energy defined by Eqs.(5.7-9)

$$\overline{(\Delta E)^2} = \overline{(H^2)} - (\bar{H})^2 \qquad (12.26)$$

The first term on the righthand side, the mean of the square of the hamiltonian, can be written in terms of the partition function according to

$$\overline{(H^2)} \equiv \frac{\int H^2 e^{-\beta H}d\lambda}{\int e^{-\beta H}d\lambda} = \frac{\partial^2\mathcal{Z}/\partial\beta^2}{\mathcal{Z}} \qquad (12.27)$$

The second term, the square of the mean value, is just

$$(\bar{H})^2 = \frac{(\partial\mathcal{Z}/\partial\beta)^2}{\mathcal{Z}^2} \qquad (12.28)$$

Thus Eq.(12.26) can be written

$$\overline{(\Delta E)^2} = \frac{\partial^2\mathcal{Z}/\partial\beta^2}{\mathcal{Z}} - \frac{(\partial\mathcal{Z}/\partial\beta)^2}{\mathcal{Z}^2} = \frac{\partial}{\partial\beta}\left(\frac{\partial\mathcal{Z}/\partial\beta}{\mathcal{Z}}\right) \qquad (12.29)$$

The mean square deviation of the energy therefore takes the very simple form

$$\overline{(\Delta E)^2} = \frac{\partial^2 ln\mathcal{Z}}{\partial\beta^2} \qquad (12.30)$$

This is a general result. Substitution of Eq.(12.24) in the present example then yields

$$\overline{(\Delta E)^2} = N\frac{\partial^2 lnz}{\partial\beta^2} \qquad (12.31)$$

The root-mean-square deviation of the energy is thus

$$(\Delta E)\text{rms} \equiv \left[\overline{(\Delta E)^2}\right]^{\frac{1}{2}} = \sqrt{N}\left(\frac{\partial^2 lnz}{\partial\beta^2}\right)^{\frac{1}{2}} \qquad (12.32)$$

It is proportional to $\sqrt{N}$ as claimed in Eq.(12.9).

The entropy for N loosely coupled subsystems is obtained by substituting Eq.(12.24) into Eq.(10.47)

$$S - S_0 = k_B N\left[lnz - \beta\frac{\partial lnz}{\partial\beta}\right] \qquad (12.33)$$

The entropy of the system is also an extensive quantity; it is proportional to the number of subsystems N.

Let us now use these arguments to characterize the order of magnitude of the various terms appearing in Eq.(12.19). We identify

$$\Delta E \cong (\Delta E)_{\text{rms}} \tag{12.34}$$

The contribution of the second and third terms in Eq.(12.19) is thus of order

$$ln\Delta E - \beta\frac{\partial ln\Delta E}{\partial\beta} \sim ln\sqrt{N} \tag{12.35}$$

In contrast, the entropy itself is an extensive quantity, and must go as N [(Eq.(12.33)]. Thus we conclude from Eq.(12.19) that

$$ln\phi(E) \sim N \tag{12.36}$$

The contribution of the terms in Eq.(12.35) is thus *completely negligible* if N is very large, and one has the approximate equality

$$S - S_0 \cong k_B ln\phi(E) \tag{12.37}$$

Now under these same conditions the logarithm of the phase extension in Eq.(12.16) yields the expression

$$ln\Omega \cong ln\phi(E) \tag{12.38}$$

Equations (12.37) and (12.38) thus imply that

$$S - S_0 \cong k_B ln\Omega \tag{12.39}$$

Hence we have derived the important relation that *the entropy of a macroscopic system is given by Boltzmann's constant times the logarithm of the phase extension.* We proceed to discuss this basic result.

The larger the value of the phase extension Ω, and hence the larger the value of $ln\Omega$ in Eq.(12.39), the larger our ignorance of just where any particular system will be in phase space. Entropy is therefore a measure of our *ignorance* because it depends on how large a phase volume one must pick in order to be "assured" that the actual system is inside that phase volume. Conversely,

$$-(S - S_0) \cong k_B ln\frac{1}{\Omega} \qquad \equiv \text{"neg-entropy"} \tag{12.40}$$

is a measure of the "amount of *information*" we have about the system. The smaller the value of the phase extension Ω, the greater this information.

Let us now consider some general principles of thermodynamics from the statistical point of view. Whereas entropy can increase or decrease in reversible processes (in fact, it remains unchanged in a reversible cycle), it always *increases* in irreversible processes. For example, if one has a thermally isolated perfect gas confined to a volume V_1 and opens a stopcock so that the gas can suddenly fill a larger volume $V_1 + V_2$ (Figure 12.3), the change in entropy is given by

$$\Delta S = k_B N ln \left(\frac{V_1 + V_2}{V_1} \right) \tag{12.41}$$

This is an elementary result of thermodynamics, and it is instructive to briefly recall its derivation. Since no heat flows and no work is done, the energy of the gas is unchanged by the first law. The energy of a perfect gas depends only on the temperature, hence T remains constant during the expansion. Now entropy is a thermodynamic function and depends only on the initial and final states. To calculate the change in entropy one merely has to find a *reversible path* between the initial and final states.

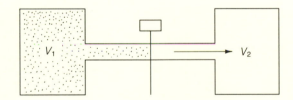

Figure 12.3 Example of the increase in entropy in an irreversible process. An ideal gas is thermally isolated in a volume V_1 and allowed to expand into a larger volume $V_1 + V_2$ without doing work.

This is readily achieved through an isothermal expansion against a piston of pressure p, the system being placed in contact with a heat bath at temperature T. The change in entropy is then given by the second and first laws

$$T dS = \delta Q = dE + p dV = p dV \tag{12.42}$$

(again, $dE = 0$ for the ideal gas under these conditions). Substitution of the equation of state (10.41) and integration between the limits V_1 and $V_1 + V_2$ yields the result in Eq.(12.41).

Consider the thermodynamic *generalization* of this result. It must be true that if some of the external parameters a_j (like the volume V) are suddenly changed and the system is left to reach a new equilibrium (without passing through intermediate equilibrium states), then the entropy must increase. To be consistent with the statistical definition of entropy in Eq.(12.39), it must be true that *the phase extension also increases in such a process* (Figure 12.4).

But what about *Liouville's Theorem*? What actually happens when a parameter a_j is suddenly changed and the phase extension then increases? Liouville's Theorem states that the *phase volume is a constant of the motion* under the new conditions (see Section 3), and thus the phase extension must be rigorously unchanged from its initial value in a "fine-grained description" of the system.†

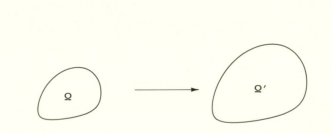

Figure 12.4 Schematic illustration of the increase in phase extension in an irreversible process.

Parts of the phase volume Ω, however, may spread out into fine filaments (Figure 12.5), and since we have already reconciled ourselves to the "coarse-grain picture" we must say that Ω has *effectively* grown larger; we are forced to say this if we are unable to distinguish the spaces between the filaments. The new system is then *effectively* indistinguishable from

† We assume for the purpose of the present discussion that the initial phase extension Ω is unmodified by the sudden change in the parameter a_j (see, for example, Fig. 12.3).

a member of the new equilibrium canonical ensemble with the new phase extension Ω'.

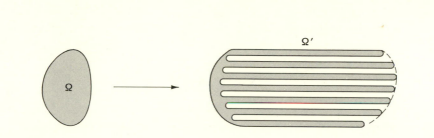

Figure 12.5 Schematic illustration of a fine-grain description of the evolution of the phase extension in a manner consistent with Liouville's Theorem.

As an instructive analogue (due to Gibbs), consider the insertion of a small ink spot in a viscous fluid such as honey, which is then stirred. The ink will spread out over the whole liquid, but when observed under a microscope, fine filaments of ink will be observed and the total volume of the ink drop has not changed (Figure 12.6). The volume of the ink is thus conserved, but upon coarse observation one would say that the ink has "spread over" the whole volume of honey.

It is also important to remember that the increase of phase extension is merely *highly probable* but not strictly necessary. Gas molecules *may* also move into a smaller volume. In this connection, we recall the form of Hamilton's equations

$$\frac{dp_i}{dt} = -\frac{\partial H}{\partial q_i} \qquad ; \qquad \frac{dq_i}{dt} = \frac{\partial H}{\partial p_i} \tag{12.43}$$

The hamiltonian of Eqs.(1.13-1.16) evidently satisfies the relation

$$H(p_i, q_i) = H(-p_i, q_i) \tag{12.44}$$

The microscopic mechanical equations are thus *invariant* under the time-reversal transformation

$$\begin{aligned} p_i &\longrightarrow -p_i \\ t &\longrightarrow -t \end{aligned} \tag{12.45}$$

Figure 12.6 Analogue of an ink drop stirred into honey.

Thus if $[p_i(t),\ q_i(t)]$ represent a set of solutions to Hamilton's equations describing a possible dynamical motion of the system

$$p_i(t),\ q_i(t) \qquad ; \quad \text{solutions to Hamilton's eqns.} \qquad (12.46)$$

then the following set of variables *also* satisfies Hamilton's equations

$$-p_i(-t),\ q_i(-t) \qquad ; \quad \text{also solutions to Hamilton's eqns.} \qquad (12.47)$$

and *also* represents a possible dynamical motion; here all the momenta are reversed and the time runs backwards.

Now it is a consequence of thermodynamic arguments that the entropy of a closed system can only increase with time. An important question is *why does the entropy increase with time when Hamilton's equations are symmetric in the time?* Equations (12.46) and (12.47) imply that there is no way to distinguish a *direction* in time merely by observing the behavior of a conservative dynamical system! One cannot tell, for example, from movies of a conservative system whether the movie runs forward or backward. This is modified by *friction*, that is, by irreversible processes. It is a fact that our notion of *"forward"* in the sense of time is due to *increasing entropy*.

Strictly speaking, a gas left to itself will expand as often as it contracts (Figure 12.7). There is no way one can assign a direction to time from an extended observation of this mechanical system. A spontaneous contraction and expansion, however, represents an enormous fluctuation; it is highly improbable and will occur only very rarely. In practice, one does not wait until a smaller volume is reached, but produces that smaller volume with external means, for example by compressing the gas with a piston. Then "after" having compressed the gas and suddenly pulling the piston back, it will indeed expand; the effect *follows* the cause.

Finally, we observe that when a constraint is removed, there is a much larger area in phase space associated with each energy surface. Since we know that the probability ρ must be constant over this surface in the new equilibrium configuration, there will be only a very small probability that the system will remain in its original equilibrium state.

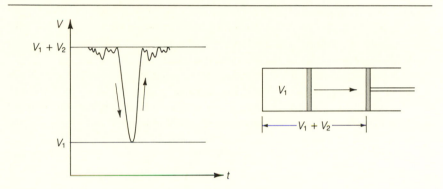

Figure 12.7 Time-dependent volume contraction and expansion of a gas in a box of volume $V_1 + V_2$. Inset illustrates experiment that produces V_1 as an initial condition.

V APPLICATIONS OF CLASSICAL STATISTICS

13. Ideal Monatomic Gases

Consider a monatomic gas with a hamiltonian given by Eqs.(1.13-1.16)

$$H = \sum_{s=1}^{N} \frac{\mathbf{p}_s^2}{2M_s} + U_I + U_W \qquad (13.1)$$

For a dilute gas the interaction potential U_I vanishes over most of phase space; we will approximate this by taking $U_I = 0$ everywhere. With this assumption, all the two-body interactions are eliminated, and the only potential that remains is that between the particles and the walls. We shall also assume that all the masses are identical, $M_s = M$.

As a first application of classical statistical mechanics we thus consider an *ideal gas* composed of a collection of mass points enclosed in a volume V. By ideal we mean that there is *no interaction between these mass points*. The hamiltonian for this system takes the form

$$H = \sum_{s=1}^{N} h_s \qquad (13.2)$$

where h_s is the energy of one of the constituent molecules

$$h_s(\mathbf{p}_s, \mathbf{r}_s) = \frac{\mathbf{p}_s^2}{2M} + U_W(\mathbf{r}_s) \qquad (13.3)$$

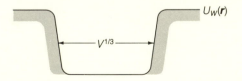

Figure 13.1 The wall potential in the single-particle hamiltonian for an ideal gas.

Here U_W is the "wall potential" sketched in Figure 13.1. We can approximate U_W by a square well of side $V^{1/3}$; it therefore satisfies the conditions

$$U_W = \begin{cases} 0 & \text{inside } V \\ \infty & \text{outside } V \end{cases} \tag{13.4}$$

In fact, we need not require that U_W is ∞ outside V, only that $U_W \gg 1/\beta = k_B T$ outside V.†

The volume element in phase space in Eq.(3.1) always factors

$$d\lambda = \prod_{s=1}^{N} d\mu_s \tag{13.5}$$

where the contribution from one of the subsystems is here defined by

$$d\mu_s \equiv dp_{sx} dp_{sy} dp_{sz} dx_s dy_s dz_s \tag{13.6}$$

We shall refer to the element of phase space for an individual subsystem as the "micro-phase space" or "μ-space." The *partition function also factors* in this problem

$$\mathcal{Z} = \int e^{-\beta H} d\lambda = \prod_{s=1}^{N} \left(\int e^{-\beta h_s} d\mu_s \right) \tag{13.7}$$
$$= z^N$$

where the *"one-particle partition function"* is defined by

$$z \equiv \int e^{-\beta h} d\mu \tag{13.8}$$

The label s can now be dropped, since it will be assumed that the subsystems are identical. Substitution of Eqs.(13.3) and (13.6) allows this partition function to be written in more detail as

$$z = \int \exp\left[-\beta(p_x^2 + p_y^2 + p_z^2)/2M\right] dp_x dp_y dp_z \int \exp\left[-\beta U_W(\mathbf{r})\right] dx dy dz \tag{13.9}$$

Because of the conditions (13.4) satisfied by the wall potential, the integrand in the final "configuration-space integral" vanishes outside of

---

† See Eqs.(13.10) and (13.38).

the confining volume and is unity inside. The configuration-space integral thus simply yields the volume V.

$$\int e^{-\beta U_W}\, dx\, dy\, dz = \int_V d^3x + \int_{\text{outside } V} e^{-\beta U_W}\, d^3x = V + 0 = V \quad (13.10)$$

Furthermore, the "momentum-space integral" itself factors, with the three cartesian components yielding identical contributions

$$\int \exp\left[-\beta(p_x^2 + p_y^2 + p_z^2)/2M\right]\, dp_x\, dp_y\, dp_z = \left(\int_{-\infty}^{\infty} e^{-\beta p^2/2M}\, dp\right)^3 \quad (13.11)$$

A change of variables in the remaining integral reduces it to dimensionless form

$$\int_{-\infty}^{\infty} e^{-\beta p^2/2M}\, dp = (2M/\beta)^{\frac{1}{2}} \int_{-\infty}^{\infty} e^{-x^2}\, dx \quad (13.12)$$

The final expression is a standard gaussian integral

$$\int_{-\infty}^{\infty} e^{-x^2}\, dx = \sqrt{\pi} \quad (13.13)$$

The one-particle partition function for the ideal gas is thus given by

$$\begin{aligned} z &= V(2\pi M/\beta)^{\frac{3}{2}} \\ &= V(2\pi M k_B T)^{\frac{3}{2}} \end{aligned} \quad (13.14)$$

Since gaussian integrals will appear repeatedly in our analysis, it is worthwhile taking a minute to analyze them. They are most conveniently done with the aid of a trick. Consider the following double integral

$$I \equiv \int_0^{\infty} dx \int_0^{\infty} dy\, \exp\left[-\lambda(x^2 + y^2)\right] = \left(\int_0^{\infty} e^{-\lambda x^2}\, dx\right)^2 \quad (13.15)$$

It can be rewritten in plane polar coordinates as an integral over the first quadrant

$$I = \int_0^{\frac{\pi}{2}} d\phi \int_0^{\infty} r\, dr\, \exp\left[-\lambda r^2\right] \quad (13.16)$$

A change of variables $u = \lambda r^2$ reduces I to an integral that is immediately evaluated.

$$I = \frac{\pi}{2} \cdot \frac{1}{2\lambda} \int_0^{\infty} e^{-u}\, du = \frac{\pi}{4\lambda} \quad (13.17)$$

The square root of this relation then yields the result

$$\int_0^\infty e^{-\lambda x^2}\, dx = \frac{1}{2}\int_{-\infty}^\infty e^{-\lambda x^2}\, dx = \frac{1}{2}\left(\frac{\pi}{\lambda}\right)^{\frac{1}{2}} \qquad (13.18)$$

All required gaussian integrals can now be evaluated by taking derivatives with respect to λ; the result in Eq.(13.13) is obtained by simply setting $\lambda = 1$.

The partition function for an ideal gas is thus given by Eqs.(13.7) and (13.14) as

$$\begin{aligned}
\mathcal{Z} &= \left[V(2\pi M/\beta)^{\frac{3}{2}}\right]^N \\
&= \left[V(2\pi M k_B T)^{\frac{3}{2}}\right]^N
\end{aligned} \qquad (13.19)$$

The logarithm of this result

$$\begin{aligned}
\ln\mathcal{Z} = N\ln z &= N\left[\ln V - \frac{3}{2}\ln\beta + \frac{3}{2}\ln(2\pi M)\right] \\
&= N\left[\ln V + \frac{3}{2}\ln T + \text{const.}\right]
\end{aligned} \qquad (13.20)$$

shows the explicit dependence on β and on the one available parameter V.

Knowledge of the partition function allows us to go to work and calculate all the thermodynamic properties of the system.

a. Energy. The energy is given by Eq.(10.7)

$$E = -\frac{\partial \ln\mathcal{Z}}{\partial \beta} = \frac{3N}{2\beta} \qquad (13.21)$$

or

$$E = \frac{3}{2}N k_B T \qquad (13.22)$$

It depends only on the temperature [verifying the discussion following Eq.(12.41)] and is independent of the volume V.

b. Heat Capacity. The constant-volume heat capacity is obtained by taking the derivative of the energy with respect to temperature at constant volume†

$$C_V = \left(\frac{\partial E}{\partial T}\right)_V = \frac{3}{2}N k_B \qquad (13.23)$$

---

† At constant volume the first law of thermodynamics states that $dE = \delta Q$.

The above expressions hold for ν_m moles; for one mole of an ideal gas one has [see Eqs.(10.43)]

$$
\begin{aligned}
E/\nu_m &= \frac{3}{2}RT \\
C_V/\nu_m &= \frac{3}{2}R \qquad ; \text{ per mole}
\end{aligned}
\tag{13.24}
$$

c. Entropy. The entropy is given by Eq.(10.47)

$$
S - S_0 = k_B \left[\ln \mathcal{Z} - \beta \frac{\partial \ln \mathcal{Z}}{\partial \beta} \right]
\tag{13.25}
$$

Hence

$$
\begin{aligned}
S - S_0 &= N k_B \left[\left(\ln V - \frac{3}{2}\ln\beta + \frac{3}{2}\ln(2\pi M) \right) - \beta \left(\frac{-3}{2\beta} \right) \right] \\
&= N k_B \left[\ln V - \frac{3}{2}\ln\beta + \text{const.} \right]
\end{aligned}
\tag{13.26}
$$

The difference in entropy between two states of an ideal gas is thus given by

$$
S_2 - S_1 = N k_B \left[\ln(V_2/V_1) + \frac{3}{2}\ln(T_2/T_1) \right]
\tag{13.27}
$$

d. Free Energy. The free energy is given by Eqs.(11.2) and (11.5)

$$
\begin{aligned}
F &= E - TS \\
&= -\frac{1}{\beta} \left[\ln \mathcal{Z} + \frac{S_0}{k_B} \right]
\end{aligned}
\tag{13.28}
$$

In a reversible process *at fixed temperature*, the change in the free energy is the work done on the system [Eq.(11.3)]; for an ideal gas this takes the form

$$
dF = \delta W = -pdV \qquad ; T = \text{const.}
\tag{13.29}
$$

Hence the *pressure* of the gas can be identified from the expression

$$
dF = -\frac{1}{\beta} \frac{\partial \ln \mathcal{Z}}{\partial V} dV = -\frac{N}{\beta V} dV
\tag{13.30}
$$

It is given by

$$
p = \frac{N}{\beta V} = \frac{N k_B T}{V}
\tag{13.31}
$$

Thus the *equation of state of an ideal gas* is

$$pV \;=\; Nk_BT \;=\; \nu_m RT \tag{13.32}$$

verifying the result anticipated in Eqs.(10.41) and (10.42).

e.   Distribution in the μ-Space.   Let us use the probability density to find the number of subsystems with momenta between $\mathbf{p}$ and $\mathbf{p} + d\mathbf{p}$ and positions between $\mathbf{r}$ and $\mathbf{r} + d\mathbf{r}$ in the equilibrium canonical ensemble. More formally, we seek the distribution function $f(\mathbf{p}, \mathbf{r})$ for finding subsystems within the volume element $d\mu$ in the μ- space. The number of subsystems lying within this volume will be defined by

$$dN \equiv f(\mathbf{p}, \mathbf{r})d\mu \tag{13.33}$$

The probability density in the canonical ensemble is given by Eqs.(9.27) and (10.6)

$$\rho \;=\; \frac{e^{-\beta H}}{\mathcal{Z}} \tag{13.34}$$

For the ideal gas this takes the form

$$\rho \;=\; \frac{1}{z^N} \prod_{s=1}^{N} e^{-\beta h_s} \tag{13.35}$$

Now the probability of finding a given subsystem, say subsystem 1, with particular values $(\mathbf{p}_1, \mathbf{r}_1) \equiv (\mathbf{p}, \mathbf{r})$ of the momenta and coordinates is obtained by integrating over all possible momenta and coordinates of the remaining $N - 1$ particles. Since the subsystems are identical, the probability of finding *a* subsystem with $(\mathbf{p}, \mathbf{r})$ is simply N times the previous result. Thus we can identify

$$f(\mathbf{p}, \mathbf{r}) = N \int \rho(\mathbf{p}, \mathbf{r}, \mathbf{p}_2, \mathbf{r}_2, \mathbf{p}_3, \mathbf{r}_3, \dots, \mathbf{p}_N, \mathbf{r}_N) \prod_{s=2}^{N} d\mu_s \tag{13.36}$$

The factorization in Eq.(13.35) allows this to be rewritten as

$$\begin{aligned}
f(\mathbf{p}, \mathbf{r}) &= \frac{Ne^{-\beta h(\mathbf{p},\mathbf{r})}}{z} \prod_{s=2}^{N} \left(\frac{1}{z} \int e^{-\beta h_s} d\mu_s \right) \\
&= \frac{Ne^{-\beta h(\mathbf{p},\mathbf{r})}}{z}
\end{aligned} \tag{13.37}$$

and thus the distribution function can be identified as

$$f(\mathbf{p}, \mathbf{r}) = \frac{N \exp\left[-\left(\frac{\mathbf{p}^2}{2M} + U_W\right)/k_B T\right]}{V(2\pi M k_B T)^{\frac{3}{2}}} \qquad (13.38)$$

With the wall potential U_W chosen according to Eq.(13.4), there will be no particles outside of the box, and inside of the box its contribution vanishes. The one-particle distribution function thus takes the form

$$
\begin{aligned}
f(\mathbf{p}, \mathbf{r}) &= \frac{N e^{-\mathbf{p}^2/2M k_B T}}{V(2\pi M k_B T)^{\frac{3}{2}}} \qquad ; \text{ inside } V \\
&= 0 \qquad\qquad\qquad\; ; \text{ outside } V
\end{aligned}
\qquad (13.39)
$$

We note that this distribution function goes as

$$f(\mathbf{p}, \mathbf{r}) \sim e^{-Mv^2/2k_B T} = e^{-(K.E.)/k_B T} \qquad (13.40)$$

It is, in fact, the celebrated *Maxwell-Boltzmann distribution* for the velocities in an ideal gas.

14. The Virial Theorem

In this section the previous analysis is used to derive a relation for the mean value in the canonical ensemble of a certain combination of dynamical variables. The resulting *virial theorem* will form a useful point of departure in discussing several applications of classical statistical mechanics.

Let $q_1, \ldots, q_n$, where $n = 3N$ is the number of degrees of freedom, be the set of generalized coordinates for the system and $p_1, \ldots, p_n$, be the set of generalized momenta. It is convenient to redefine these variables according to

$$
\begin{aligned}
p_1, \ldots, p_n &\equiv \xi_1, \ldots, \xi_n \\
q_1, \ldots, q_n &\equiv \xi_{n+1}, \ldots, \xi_{2n} \quad ; \; n = 3N
\end{aligned}
\qquad (14.1)
$$

In this fashion position and momenta are no longer distinguished, and all dynamical variables are treated on a uniform basis; they are here denoted by $\{\xi_k; \; k = 1, 2 \ldots, 2n\}$. The hamiltonian evidently has the functional form

$$H = H(\xi_1, \ldots, \xi_{2n}) \qquad (14.2)$$

In this same notation, the volume element in phase space is given by

$$d\lambda = \prod_{i=1}^{n} dp_i dq_i = \prod_{k=1}^{2n} d\xi_k \tag{14.3}$$

Consider the mean value of the quantity $\xi_k \partial H / \partial \xi_k$ defined by the expression

$$\overline{\left(\xi_k \frac{\partial H}{\partial \xi_k} \right)} = \frac{\int \xi_k \frac{\partial H}{\partial \xi_k} e^{-\beta H} d\lambda}{\int e^{-\beta H} d\lambda} \tag{14.4}$$

This relation holds for the kth coordinate; there is no implied summation on k. The identity

$$\frac{\partial H}{\partial \xi_k} e^{-\beta H} = -\frac{1}{\beta} \frac{\partial}{\partial \xi_k} e^{-\beta H} \tag{14.5}$$

can be used to rewrite Eq.(14.4)

$$\overline{\left(\xi_k \frac{\partial H}{\partial \xi_k} \right)} = \frac{-\frac{1}{\beta} \int \xi_k \frac{\partial}{\partial \xi_k} e^{-\beta H} d\lambda}{\int e^{-\beta H} d\lambda} \tag{14.6}$$

Now assume that $H \longrightarrow \infty$ as $|\xi_k| \longrightarrow \infty$. This is true, for example, for the hamiltonian in Eqs.(1.13-1.16) describing interacting particles in a container; it is a positive definite quadratic form in the momenta, and the wall potential U_W is here defined to be infinitely large for values of the particle coordinates that lie outside of the box as in Eq.(13.4). The factor $\exp(-\beta H)$ in the integrand then provides exponential damping, and one can evaluate the numerator in Eq.(14.6) by partial integration. This leads to the expression

$$\overline{\left(\xi_k \frac{\partial H}{\partial \xi_k} \right)} = \frac{\frac{1}{\beta} \int e^{-\beta H} d\lambda}{\int e^{-\beta H} d\lambda} \tag{14.7}$$

But the remaining ratio is just unity, *independent of the details of the dynamical interactions*. We thus arrive at the virial theorem

$$\overline{\left(\xi_k \frac{\partial H}{\partial \xi_k} \right)} = \frac{1}{\beta} = k_B T \tag{14.8}$$

In words this relation states that if one takes any one of the $6N$ "coordinates" ξ_k in Eq.(14.1), then the mean value of the product of this coordinate and the derivative of the hamiltonian with respect to this coordinate is given by $1/\beta = k_B T$ independent of the details of dynamical interactions. This follows from the nature of the canonical ensemble, the only additional assumption being that the expression $\exp(-\beta H)$ must vanish when the coordinate ξ_k gets very large.

As an application of the virial theorem consider the behavior of a gas of point particles. The hamiltonian is given by

$$H = T + U$$

$$T = \sum_{s=1}^{N} \frac{\mathbf{p}_s^2}{2M_s} \tag{14.9}$$

$$U = U(\mathbf{r}_1, \ldots, \mathbf{r}_N)$$

If one first takes ξ_k to be the x component of the momentum of the sth particle, then

$$\xi_k = p_{sx}$$

$$\frac{\partial H}{\partial \xi_k} = \frac{p_{sx}}{M_s} \tag{14.10}$$

The virial theorem states that the mean value of the product of these quantities is given by

$$\overline{\left(\xi_k \frac{\partial H}{\partial \xi_k}\right)} = \overline{\left(\frac{p_{sx}^2}{M_s}\right)} = k_B T \tag{14.11}$$

Since the same result holds for each of the other two cartesian components of the momentum, one has

$$\overline{\left(\frac{\mathbf{p}_s^2}{M_s}\right)} = 3k_B T \tag{14.12}$$

and hence the mean kinetic energy of the sth particle is given by

$$\overline{\left(\frac{\mathbf{p}_s^2}{2M_s}\right)} = \frac{3}{2}k_B T \tag{14.13}$$

The mean kinetic energy for the entire gas is obtained by summing over all particles

$$\overline{\left(\sum_{s=1}^{N} \frac{\mathbf{p}_s^2}{2M_s}\right)} = \sum_{s=1}^{N} \overline{\left(\frac{\mathbf{p}_s^2}{2M_s}\right)} = \frac{3}{2}Nk_B T \tag{14.14}$$

We shall denote this mean kinetic energy of the gas by $\overline{(K.E.)}$, and therefore

$$\overline{(K.E.)} = \frac{3}{2}Nk_B T \tag{14.15}$$

Equation (14.15) is a general result arising from an application of the virial theorem to the hamiltonian in Eq.(14.9). For an *ideal gas* there are no interparticle potentials, and hence Eq.(14.15) also gives the total energy of the confined system in this case

$$E = \overline{(K.E.)} = \frac{3}{2}Nk_BT \quad ; \text{ ideal gas} \tag{14.16}$$

Next take the coordinate ξ_k to be the x component of the position of the sth particle. The derivative $\partial H/\partial \xi_k$, evaluated from Eq.(14.9), is

$$\frac{\partial H}{\partial \xi_k} = \frac{\partial U}{\partial x_s} = -F_{sx} \tag{14.17}$$

The final equality follows, since the negative gradient of the potential is the force. The virial theorem gives the result

$$\overline{\left(\xi_k \frac{\partial H}{\partial \xi_k}\right)} = -\overline{(x_s F_{sx})} = k_B T \tag{14.18}$$

Since the same relation holds for the other two cartesian components of the position, one has, in vector notation,

$$-\overline{(\mathbf{r}_s \cdot \mathbf{F}_s)} = 3k_B T \tag{14.19}$$

We are thus motivated to define a term called the "virial" that applies to the whole system

$$\text{virial} \equiv \mathcal{V} \equiv -\overline{\left(\sum_{s=1}^{N} \mathbf{r}_s \cdot \mathbf{F}_s\right)}$$

$$\mathcal{V} = -\sum_{s=1}^{N} \overline{(\mathbf{r}_s \cdot \mathbf{F}_s)} = 3Nk_B T \tag{14.20}$$

The last equality follows from the virial theorem discussed above.

Let us first evaluate the virial for the ideal monatomic gas confined to a box of side L. The only forces in this problem are due to the walls, as illustrated in Figure 14.1. When the particle strikes the wall, it feels an inward force pushing it into the box of magnitude

$$F_x = -\frac{\partial}{\partial x}U_W \tag{14.21}$$

Since F_x is non-vanishing only for $x = 0$ and $x = L$, it is very simple to compute the product of (position)×(force) that occurs in the virial. The contribution of the x component to the virial in Eq.(14.20) is given by

$$-\sum_{s=1}^{N} \overline{(x_s F_{sx})} = -L \sum_{s=1}^{N} \bar{F}_{sx}|_{x=L} \qquad (14.22)$$

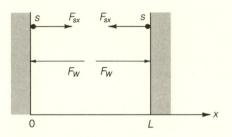

Figure 14.1 The x component of the forces on the sth particle of an ideal monatomic gas confined to a box of volume L^3 used in evaluating the "virial" for this system. F_W is the total reaction force on the wall.

The virial theorem implies that this expression is just $N k_B T$

$$-L \sum_{s=1}^{N} \bar{F}_{sx}|_{x=L} = N k_B T \qquad (14.23)$$

By Newton's third law, the force that the particles exert on the wall must be equal and opposite to the force the wall exerts on the particles

$$-\sum_{s=1}^{N} \bar{F}_{sx}|_{x=L} = F_W = L^2 p \qquad (14.24)$$

The second equality follows from the definition of pressure as force/area. Equations (14.23) and (14.24) can be combined to give

$$L^3 p = N k_B T \qquad (14.25)$$

which provides an alternate derivation of the ideal gas law

$$pV = Nk_BT \quad ; \text{ideal gas} \tag{14.26}$$

The total virial for an ideal gas in Eq.(14.20) is 3 times the value obtained from the x component in Eq.(14.22), since there is a similar contribution from the other two pairs of walls in the box.

$$\mathcal{V}_1 = 3pV = 3Nk_BT \tag{14.27}$$

Now the reader may well feel that the use of the virial theorem is a cumbersome way to rederive the properties of an ideal gas. Its utility lies in the fact that it provides a framework for discussing the behavior of the *non-ideal gas* where the dynamics is governed by a hamiltonian containing interparticle interactions. We shall, in fact, use the virial theorem to derive the virial expansion of the equation of state of the non-ideal gas.

To proceed, we include the term U_I of Eq.(13.1) in Sec. 13, which was there discarded on the grounds of vanishingly small density. Here the leading corrections at finite density will be evaluated. The interaction term will be assumed to be composed of two-body potentials as in Eq.(1.16). We define (see Figure 14.2)

$$\mathbf{F}_{ss'} = \text{force on particle } s \text{ due to } s' \tag{14.28}$$

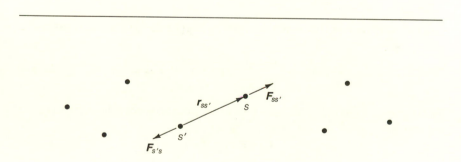

Figure 14.2 Two-body force used in calculating second contribution to the virial in an interacting monatomic gas.

The total force on particle s arising from all of the other particles is thus given by

$$\mathbf{F}_s^T = \sum_{s'=1}^{N} \mathbf{F}_{ss'} \quad ; \mathbf{F}_{ss} \equiv 0 \tag{14.29}$$

Here we define $\mathbf{F}_{ss} \equiv 0$; there is no self-force, and there is no self-energy in the hamiltonian. For a non-ideal gas, the virial will thus have a second contribution

$$\mathcal{V}_2 = -\sum_{s=1}^{N} \overline{(\mathbf{r}_s \cdot \mathbf{F}_s^T)} = -\sum_{s=1}^{N}\sum_{s'=1}^{N} \overline{(\mathbf{r}_s \cdot \mathbf{F}_{ss'})} \qquad (14.30)$$

Newton's third law can again be invoked; it states that

$$\mathbf{F}_{ss'} = -\mathbf{F}_{s's} \qquad (14.31)$$

Hence the second contribution to the virial can also be written as

$$\mathcal{V}_2 = +\sum_{s=1}^{N}\sum_{s'=1}^{N} \overline{(\mathbf{r}_s \cdot \mathbf{F}_{s's})} \qquad (14.32)$$

An interchange of dummy summation variables $s \leftrightarrow s'$ yields †

$$\mathcal{V}_2 = +\sum_{s=1}^{N}\sum_{s'=1}^{N} \overline{(\mathbf{r}_{s'} \cdot \mathbf{F}_{ss'})} \qquad (14.33)$$

and the addition of this result to that in Eq.(14.30) gives the expression

$$2\mathcal{V}_2 = -\sum_{s=1}^{N}\sum_{s'=1}^{N} \overline{[(\mathbf{r}_s - \mathbf{r}_{s'}) \cdot \mathbf{F}_{ss'}]} \qquad (14.34)$$

With the introduction of the following notation for the interparticle separation

$$\mathbf{r}_{ss'} \equiv \mathbf{r}_s - \mathbf{r}_{s'} \qquad (14.35)$$

the second contribution to the virial arising from pairwise interactions takes the compact form

$$\mathcal{V}_2 = -\frac{1}{2}\sum_{s=1}^{N}\sum_{s'=1}^{N} \overline{(\mathbf{r}_{ss'} \cdot \mathbf{F}_{ss'})} \qquad (14.36)$$

We shall assume that the interaction depends only on the magnitude of the distance between the particles

$$U(|\mathbf{r}_s - \mathbf{r}_{s'}|) = U(r_{ss'}) \qquad (14.37)$$

---

† Note that the order of finite sums can always be interchanged.

In this case the force is given by

$$\mathbf{F}_{ss'} = -\nabla_s U(|\mathbf{r}_s - \mathbf{r}_{s'}|) \tag{14.38}$$

The corresponding term in the virial

$$\overline{[\mathbf{r}_{ss'} \cdot \nabla_s U(|\mathbf{r}_s - \mathbf{r}_{s'}|)]} = \overline{[\mathbf{r}_{s's} \cdot \nabla_{s'} U(|\mathbf{r}_s - \mathbf{r}_{s'}|)]} \tag{14.39}$$

is then explicitly symmetric in s and s'. Thus if we now specialize to the case of identical particles and define

$$\overline{[\mathbf{r}_{ss'} \cdot \nabla_s U(r_{ss'})]} \equiv \overline{[\mathbf{r} \cdot \nabla U(r)]} \tag{14.40}$$

then the second contribution to the virial receives an identical contribution from each of the possible pairs and is given by

$$
\begin{aligned}
\mathcal{V}_2 &= \frac{1}{2}N(N-1)\overline{[\mathbf{r} \cdot \nabla U(r)]} \\
&\cong \frac{1}{2}N^2\overline{[\mathbf{r} \cdot \nabla U(r)]} \qquad ; \text{ identical particles}
\end{aligned} \tag{14.41}
$$

We proceed to evaluate this expression, which involves the mean value of the product of the interparticle separation and the gradient of the potential for a pair of interacting particles in the gas.

The probability density in the canonical ensemble is given by Eq.(9.27)

$$\rho = \frac{e^{-\beta H}}{\int e^{-\beta H} d\lambda} \tag{14.42}$$

and this is to be used in evaluating the mean value of the two-particle interaction appearing in Eq.(14.41)

$$\overline{[\mathbf{r} \cdot \nabla U(r)]} = \frac{\int \mathbf{r}_{ss'} \cdot \nabla_s U(r_{ss'})e^{-\beta H} d\lambda}{\int e^{-\beta H} d\lambda} \tag{14.43}$$

Since the expression whose mean value we seek does not involve momenta, all momentum integrals in the numerator on the righthand side will factor and will be cancelled by identical factors appearing in the denominator. The mean value will therefore only involve a ratio of *configuration space integrals*. These are complicated $3N$-dimensional multiple integrals in the general case; however, if the system is at sufficiently *low density*, then the only interaction term in the weighting factor that will be non-zero within the range of the two-body potential $U(r_{ss'})$ will be that coming from this potential itself, all other particles and the walls being sufficiently far away so that their interaction potentials will have effectively vanished at the position $r_{s'}$. Furthermore, the relevant configuration-space volume element can be rewritten as

$$dr_{s'} dr_s = dr_{s'} d(r_s - r_{s'}) = dr_{s'} dr_{ss'} \tag{14.44}$$

where one first integrates over all *relative* coordinates $r_{ss'} = r_s - r_{s'}$, keeping the position of the s'^{th} particle fixed, and then finally integrates over $r_{s'}$. The only relevant integral is now the integral over the relative coordinate. All other configuration-space integrals will *factor* in the low-density approximation and again cancel in the ratio on the righthand side of Eq.(14.43). Thus, if the interacting gas is at sufficiently low density, the mean value in Eq.(14.43) is evaluated by the following ratio of integrals

$$\overline{[r \cdot \nabla U(r)]} = \frac{\int r \cdot \nabla U(r) e^{-\beta U(r)} dr}{\int e^{-\beta U(r)} dr} \tag{14.45}$$

The effective probability now appearing on the righthand side of this expression

$$P_2(r) dr \equiv \frac{e^{-\beta U(r)} dr}{\int e^{-\beta U(r)} dr} \tag{14.46}$$

has a simple interpretation. It is the probability of finding a second atom within a volume element of atomic dimensions surrounding a first atom. The assumption of low density for the interacting system implies that the only relevant factor remaining from the full probability density in Eq.(14.42) is that coming from the two-body potential $U(r)$ itself, and the probability density in Eq.(14.46) again has a simple structure given by a *Boltzmann distribution* [compare Eq.(13.40)]

$$P_2(r) \sim e^{-U(r)/k_B T} = e^{-(P.E.)/k_B T} \tag{14.47}$$

A schematic representation of $U(r)$ for a real gas is sketched in Figure 14.3. It arises from the "Van der Waal's forces" between the atoms.

The region over which $U(r) \neq 0$ is typically a few angstroms. For a macroscopic system, the denominator in Eq.(14.45) is therefore effectively the total volume

$$\int_V e^{-\beta U} dr \cong V \tag{14.48}$$

and thus

$$\overline{[r \cdot \nabla U(r)]} = \overline{\left(r \frac{dU}{dr}\right)} = \frac{1}{V} \int r \frac{dU}{dr} e^{-\beta U} dr \tag{14.49}$$

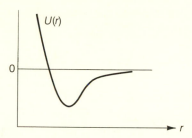

Figure 14.3 Schematic illustration of the two-body interaction potential for a real gas.

The second contribution to the virial in Eq.(14.41) for a low-density system of identical monatomic particles has thus been reduced to the evaluation of the following one-dimensional integral

$$\mathcal{V}_2 = \frac{1}{2}N^2 \cdot \frac{4\pi}{V} \int_0^\infty r^3 \frac{dU}{dr} e^{-\beta U} dr \qquad (14.50)$$

This result is still fairly general, however, since only the approximate range of the interaction potential has been used in its derivation.

The total virial for the interacting gas is now the sum of the contribution in Eq.(14.27) coming from the walls and the term in Eq.(14.50) coming from the two-particle interactions. According to the virial theorem of Eq.(14.20), the total virial is just $3Nk_BT$. Thus we have

$$\mathcal{V}_1 + \mathcal{V}_2 = 3pV + \mathcal{V}_2 = 3Nk_BT \qquad (14.51)$$

This expression yields the following equation of state for the non-ideal gas

$$pV = Nk_BT - \frac{1}{3}\mathcal{V}_2 = Nk_BT\left(1 - \frac{\mathcal{V}_2}{3Nk_BT}\right)$$

$$pV \equiv Nk_BT\left(1 - \frac{\tilde{b}_2}{V}\right) \qquad (14.52)$$

The term $\tilde{b}_2/V$ represents a correction to the perfect gas law arising from the Van der Waal's forces. It defines the *second virial coefficient*

$$\frac{\mathcal{V}_2}{3Nk_BT} \equiv \frac{\tilde{b}_2}{V} \quad ; \ \tilde{b}_2 \equiv \text{second virial coefficient} \qquad (14.53)$$

If one includes more general terms for $3, 4, \ldots, N$-body correlations in the configuration-space integral discussed below Eq.(14.43), then the equation of state becomes a power series in $1/V$ containing higher virial coefficients.

Let us then proceed to evaluate the second virial $\mathcal{V}_2$ in Eq.(14.50). The use of the following identity

$$\frac{dU}{dr}e^{-\beta U} = -\frac{1}{\beta}\frac{d}{dr}e^{-\beta U} = -\frac{1}{\beta}\frac{d}{dr}(e^{-\beta U} - 1) \tag{14.54}$$

allows us to rewrite the integral

$$\mathcal{V}_2 = \frac{2\pi N^2}{V}\int_0^\infty r^3(-\frac{1}{\beta})\frac{d}{dr}(e^{-\beta U} - 1)dr \tag{14.55}$$

The advantage of this form is that the factor $[\exp(-\beta U) - 1]$ only fails to vanish within the range of the two-body interaction; it is zero over most of the volume, and its presence allows one to carry out partial integrations at will. For example, a single partial integration of Eq.(14.55) yields

$$\mathcal{V}_2 = \frac{2\pi N^2}{\beta V}\int_0^\infty 3r^2(e^{-\beta U} - 1)dr \tag{14.56}$$

Evidently the boundary terms vanish in this partial integration

$$r^3(e^{-\beta U} - 1)|_0^\infty = 0 \tag{14.57}$$

if the two-body potential $U \longrightarrow 0$ fast enough at large interparticle separation $r \longrightarrow \infty$. Equations (14.56) and (14.53) imply that the second virial coefficient is given by the expression

$$\tilde{b}_2 = 2\pi N\int_0^\infty r^2(e^{-\beta U} - 1)dr \tag{14.58}$$

This is an important result, since it relates the leading finite-density modification of the equation of state of the non-ideal gas in Eq.(14.52) to an integral over the two-particle interaction potential between the subsystems.

Let us compare these results with the familiar Van der Waal's equation of state given by

$$\left(p + \frac{a}{V^2}\right)(V - b) = Nk_BT \tag{14.59}$$

where a and b are two constants characterizing the non-ideal gas. This equation of state can be rewritten in the form

$$pV = \frac{Nk_BT}{(1 - b/V)} - \frac{a}{V} \cong Nk_BT\left(1 + \frac{b}{V} - \frac{a}{Nk_BTV}\right) \tag{14.60}$$

Now consider the model "Van der Waal's potential" sketched in Figure 14.4, which satisfies the conditions†

$$U = \infty \qquad\qquad 0 \leq r \leq r_1$$
$$|U| << k_B T \qquad r_1 < r \leq \infty \tag{14.61}$$

In this case, the integrand appearing in Eq.(14.58) can be written as

$$(e^{-\beta U} - 1) = -1 \qquad\qquad 0 \leq r \leq r_1$$
$$\cong -\beta U \qquad\qquad r_1 < r \leq \infty \tag{14.62}$$

and the second virial coefficient becomes

$$\tilde{b}_2 \cong -2\pi N \int_0^{r_1} r^2 dr - \frac{2\pi N}{k_B T} \int_{r_1}^{\infty} r^2 U(r) dr \tag{14.63}$$

Note that the first term in this expression is negative for the potential sketched in Figure 14.4, while the second term is positive. The equation of state in Eq.(14.52) thus takes the form

$$pV = N k_B T \left[1 + \frac{2\pi N r_1^3}{3V} + \frac{2\pi N}{k_B T V} \int_{r_1}^{\infty} r^2 U(r) dr \right] \tag{14.64}$$

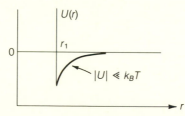

Figure 14.4 Van der Waal's potential used in evaluation of the second virial coefficient.

† Recall that $k_B T \cong (1/40)eV$ at $T = 300° K$.

A comparison with the *Van der Waal's equation of state* in Eq.(14.60) allows an identification of the two constants a and b in terms of properties of the two-body potential in Figure 14.4

$$b = \frac{2\pi N}{3}r_1^3$$

$$a = -2\pi N^2 \int_{r_1}^{\infty} r^2 U(r)dr$$

(14.65)

Note that b is positive and is a measure of the size of the hard-core repulsive interaction region; a is also positive and is a measure of the integrated strength of the attractive tail.

15. The Equipartition Theorem

An important application of the virial theorem is the *equipartition theorem*. We use the notation of Eq.(14.1), where the generalized coordinates and momenta are treated on the same footing, and the dynamical variables are denoted by $\{\xi_k; \ k = 1, 2, \ldots, 2n\}$. Consider a hamiltonian

$$H(p_i, q_i) = H(\xi_k)$$

(15.1)

which is a homogeneous function of the νth degree in the coordinates $\xi_1, \xi_2, \ldots, \xi_g$ with an arbitrary dependence on the remaining coordinates $\xi_{g+1}, \xi_{g+2}, \ldots, \xi_{2n}$. The analytic representation of this assumed structure is

$$H(\lambda\xi_1, \lambda\xi_2, \ldots, \lambda\xi_g, \xi_{g+1}, \ldots, \xi_{2n}) = \lambda^\nu H(\xi_k)$$

(15.2)

where λ is an arbitrary scale factor. One example of a hamiltonian possessing this property is the quadratic form

$$H = \sum_{k=1}^{g} \sum_{l=1}^{g} a_{kl}(\xi_{g+1}, \ldots, \xi_{2n})\xi_k\xi_l$$

(15.3)

with the coefficients a_{kl} being arbitrary functions of $\xi_{g+1}, \ldots, \xi_{2n}$. The expression in Eq.(15.3) is said to be a homogeneous function of the second degree in $\xi_1, \ldots, \xi_g$. Another example is

$$H = (a\xi_1^2 + b\xi_2^2)^{\frac{1}{2}}$$

(15.4)

which is a homogeneous function of first degree in ξ_1, ξ_2 ; a homogeneous function does not have to be a polynomial.

Now if a function $f(\xi_1, \ldots, \xi_g)$ is homogeneous of the νth degree

$$f(\lambda \xi_1, \lambda \xi_2, \ldots, \lambda \xi_g) = \lambda^\nu f(\xi_1, \ldots, \xi_g) \tag{15.5}$$

then it satisfies Euler's Theorem

$$\sum_{k=1}^{g} \xi_k \frac{\partial f}{\partial \xi_k} = \nu f \tag{15.6}$$

The proof follows immediately upon taking the derivative of Eq.(15.5) with respect to λ and then setting $\lambda = 1$.

$$\frac{d}{d\lambda} f(\lambda \xi_1, \ldots, \lambda \xi_g)|_{\lambda=1} = \sum_{k=1}^{g} \xi_k \frac{\partial f}{\partial \xi_k}$$

$$= \nu \lambda^{\nu-1} f(\xi_1, \ldots, \xi_g)|_{\lambda=1} = \nu f \tag{15.7}$$

Application of Euler's Theorem to the hamiltonian in Eq.(15.2) leads to the result

$$\sum_{k=1}^{g} \xi_k \frac{\partial H}{\partial \xi_k} = \nu H \tag{15.8}$$

We are now in a position to apply the virial theorem of Eq.(14.8), which states that

$$\overline{\left(\xi_k \frac{\partial H}{\partial \xi_k} \right)} = k_B T \tag{15.9}$$

Recall that the only assumption entering into the derivation of this result is that the expression $\exp(-\beta H)$ must vanish when the coordinate ξ_k gets very large; in particular, the quadratic form in the example in Eq.(15.3) must be positive definite. If this holds, then the application of the virial theorem to the expression in Eq.(15.8) gives the result

$$\overline{\left(\sum_{k=1}^{g} \xi_k \frac{\partial H}{\partial \xi_k} \right)} = \sum_{k=1}^{g} \overline{\left(\xi_k \frac{\partial H}{\partial \xi_k} \right)} = \nu \bar{H} \tag{15.10}$$

Since the mean value of the hamiltonian is just the energy, we have

$$g k_B T = \nu E \tag{15.11}$$

or

$$E = g \frac{k_B T}{\nu} \tag{15.12}$$

This is the general statement of the equipartition theorem. For the special case of a quadratic form with $\nu = 2$

$$E = g\frac{k_B T}{2} \qquad ; \text{ quadratic form} \qquad (15.13)$$

In words this result states: "If the hamiltonian is a positive definite quadratic form in a certain number (g) of dynamical variables (be they $p's$ or $q's$), each of them contributes the amount $k_B T/2$ to the energy." This result is sometimes loosely stated: "Each degree of freedom contributes $k_B T/2$ to the energy"; however, we have derived the precise form of the equipartition theorem.

This analysis can be specialized slightly to a form that will be useful in several applications. Assume that the hamiltonian is a finite sum of contributions

$$H = \sum_j H_j(\xi_k) \qquad (15.14)$$

Assume further that one term, say H_i, is a homogeneous function of νth degree in the coordinates $\xi_1, \ldots, \xi_g$ and that these coordinates appear only in the term H_i. Thus

$$H_i(\lambda\xi_1, \lambda\xi_2, \ldots, \lambda\xi_g, \xi_{g+1}, \ldots, \xi_{2n}) = \lambda^\nu H_i(\xi_k) \qquad (15.15)$$

and Euler's Theorem now gives the result

$$\sum_{k=1}^{g} \xi_k \frac{\partial H_i}{\partial \xi_k} = \sum_{k=1}^{g} \xi_k \frac{\partial H}{\partial \xi_k} = \nu H_i(\xi_k) \qquad (15.16)$$

The application of the virial theorem to the mean value of this expression gives

$$\overline{\left(\sum_{k=1}^{g} \xi_k \frac{\partial H}{\partial \xi_k} \right)} = g k_B T = \nu \bar{H}_i \qquad (15.17)$$

and hence

$$E_i = g\frac{k_B T}{\nu} \qquad (15.18)$$

Thus this term in the hamiltonian by itself makes a contribution to the total energy identical in form to that given in Eq.(15.12); there is a similar contribution to the energy for each term in the hamiltonian that has these properties.

Suppose further that the set of generalized coordinates and momenta is enlarged by including dynamical variables describing the internal structure of the subsystems. The application to diatomic molecules in Sec. 16 represents just such an extension. The set of dynamical variables can again be described by $\{\xi_k; \ k = 1, 2 \ldots, 2n\}$, where n is now the total number of degrees of freedom of the system, and the results obtained in Eqs.(15.12) and (15.18) again apply.

16. Examples

In this section we consider the application of classical statistical mechanics to several fundamental physical systems.

a) Mean Kinetic Energy. Consider once again the example of the ideal monatomic gas with hamiltonian

$$H = \sum_{s=1}^{N} \left[\frac{1}{2m}(p_{sx}^2 + p_{sy}^2 + p_{sz}^2) + U_W(\mathbf{r}_s) \right] \qquad (16.1a)$$

If we suppress the wall potential, which merely serves to confine the gas to a container of volume V and does not contribute to the total internal energy of the system, then the hamiltonian

$$H = \sum_{s=1}^{N} \frac{1}{2m}(p_{sx}^2 + p_{sy}^2 + p_{sz}^2) \qquad (16.2a)$$

is observed to be a positive definite quadratic form in the $g = 3N$ momentum coordinates. The equipartition theorem of Eq.(15.13) immediately yields the energy of the gas, which in this case is just its mean kinetic energy

$$E = \frac{3}{2}Nk_BT \qquad (16.3a)$$

This expression is identical to that given in Eq.(13.22).

b) Diatomic Molecules. Consider next an ideal gas of diatomic molecules, whose internal structure will be modelled by two atoms of mass m_1 and m_2 moving in an interatomic potential $U(|\mathbf{r}_1 - \mathbf{r}_2|)$. First, introduce the relative and center-of-mass coordinates (Figure 16.1b)

$$\begin{aligned} \mathbf{r} &\equiv \mathbf{r}_1 - \mathbf{r}_2 \\ \mathbf{R} &\equiv \frac{m_1\mathbf{r}_1 + m_2\mathbf{r}_2}{m_1 + m_2} \end{aligned} \qquad (16.1b)$$

and the relative and center-of-mass momenta

$$\begin{aligned} \mathbf{p} &\equiv \frac{m_2\mathbf{p}_1 - m_1\mathbf{p}_2}{m_1 + m_2} \\ \mathbf{P} &\equiv \mathbf{p}_1 + \mathbf{p}_2 \end{aligned} \qquad (16.2b)$$

The hamiltonian for this system is given by

$$h = \frac{\mathbf{p}_1^2}{2m_1} + \frac{\mathbf{p}_2^2}{2m_2} + U(|\mathbf{r}_1 - \mathbf{r}_2|) + U_W(\mathbf{R}) \qquad (16.3b)$$

where the interatomic potential U(r) is schematically illustrated in Figure 16.2b.

It is an elementary exercise in mechanics to show that this hamiltonian can be rewritten in terms of the center-of-mass and relative momenta according to

$$h = \frac{\mathbf{P}^2}{2M} + \frac{\mathbf{p}^2}{2\mu} + U(r) + U_W(\mathbf{R}) \qquad (16.4b)$$

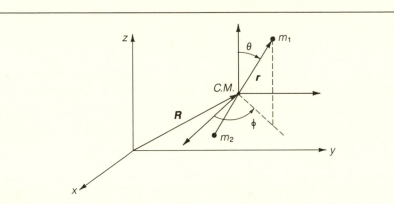

Figure 16.1b The center-of-mass coordinate system for a diatomic molecule.

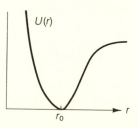

Figure 16.2b Schematic illustration of the interatomic potential $U(r)$ in a diatomic molecule.

where the total and reduced mass are defined by

$$M \equiv m_1 + m_2$$
$$\mu \equiv \frac{m_1 m_2}{m_1 + m_2} \tag{16.5b}$$

It is readily established (Problem 16.b6) that the jacobian for the transformation in Eqs.(16.1b) and (16.2b) is unity (the transformation is, in fact, canonical), and hence the volume in micro-phase space can be written in terms of center-of-mass and relative coordinates according to

$$d\mathbf{p}_1 d\mathbf{p}_2 d\mathbf{r}_1 d\mathbf{r}_2 = d\mathbf{P} d\mathbf{p} d\mathbf{R} d\mathbf{r} \tag{16.6b}$$

The center-of-mass and relative coordinates and momenta can now be taken as the new generalized dynamical variables for this system.

The result in Eq.(1.56) allows the internal energy of the molecule to be rewritten further, in terms of the spherical coordinates in Figure 16.1b, as

$$h_{\text{int}} \equiv \frac{\mathbf{p}^2}{2\mu} + U(r)$$
$$= \frac{1}{2\mu} \left(p_r^2 + \frac{p_\theta^2}{r^2} + \frac{p_\phi^2}{r^2 \sin^2 \theta} \right) + U(r) \tag{16.7b}$$

The new variable set $\{p_r, p_\theta, p_\phi, r, \theta, \phi\}$ was derived in Sec. 1 through a canonical transformation; therefore the volume element in phase space is again preserved (Appendix B) as

$$d\mathbf{p} d\mathbf{r} = dp_r dp_\theta dp_\phi dr d\theta d\phi \tag{16.8b}$$

The total hamiltonian for the ideal diatomic gas is the sum of the contributions from each of the N molecules

$$H = \sum_{s=1}^{N} h_s \tag{16.9b}$$

The molecules will be assumed to be identical with individual hamiltonians given by Eq.(16.7b)

$$h = \frac{\mathbf{P}^2}{2M} + \frac{1}{2\mu} \left(p_r^2 + \frac{p_\theta^2}{r^2} + \frac{p_\phi^2}{r^2 \sin^2 \theta} \right) + U(r) + U_W(\mathbf{R})$$
$$;\quad \text{each } s \tag{16.10b}$$

The volume element in phase space factors

$$d\lambda = \prod_{s=1}^{N} d\mu_s \qquad (16.11b)$$

and the volume element in the μ-space is now given by Eqs.(16.6b) and (16.8b)

$$d\mu = d\mathbf{P}d\mathbf{R}dp_r\,dp_\theta dp_\phi dr d\theta d\phi \qquad ; \text{ each } s \qquad (16.12b)$$

The partition function for the system therefore also factors in this problem exactly as in Eq.(13.7)

$$\mathcal{Z} = z^N \qquad (16.13b)$$

where

$$z \equiv \int d\mathbf{P}d\mathbf{R}dp_r\,dp_\theta dp_\phi dr d\theta d\phi \,\exp(-\beta h) \qquad (16.14b)$$

The integral over the center-of-mass coordinates is carried out exactly as in Sec. 13 with the result that [Eq.(13.14)]

$$z = V(2\pi M k_B T)^{\frac{3}{2}} z_{\text{int}} \qquad (16.15b)$$

The remaining *internal partition function* is defined by

$$z_{\text{int}} \equiv \int dp_r\,dp_\theta dp_\phi dr d\theta d\phi \,\exp(-\beta h_{\text{int}}) \qquad (16.16b)$$

with h_{int} given by Eq.(16.7b).

Rigid Rotation. If the minimum in the interatomic potential in Figure 16.2b is sufficiently sharp, then the radius appearing in the rotational part of the internal hamiltonian can be replaced by its equilibrium value $r = r_0$. The molecule in Figure 16.1b tumbles like a dumbbell of well-defined interatomic separation, and the internal hamiltonian of Eq.(16.7b) can be approximated as

$$h_{\text{int}} \cong \frac{1}{2\mu}\left(p_r^2 + \frac{p_\theta^2}{r_0^2} + \frac{p_\phi^2}{r_0^2\sin^2\theta}\right) + U(r) \qquad ; \text{ rigid rotation} \quad (16.17b)$$

The internal hamiltonian now becomes a sum of independent angular and radial terms, and since the remaining μ-phase space volume is a product, the internal partition function in Eq.(16.16b) also *factors*

$$z_{\text{int}} \cong z_{\text{rot}} \cdot z_{\text{vib}} \qquad (16.18b)$$

The rotation and vibration contributions are here defined by

$$z_{\text{rot}} \equiv \int dp_\theta dp_\phi d\theta d\phi \ \exp\left\{-\beta\left[\frac{1}{2\mu r_0^2}\left(p_\theta^2 + \frac{p_\phi^2}{\sin^2\theta}\right)\right]\right\} \qquad (16.19b)$$

$$z_{\text{vib}} \equiv \int dp_r dr \ \exp\left\{-\beta\left[\frac{p_r^2}{2\mu} + U(r)\right]\right\} \qquad (16.20b)$$

The logarithm of the partition function for the system, from which its thermodynamic properties can be calculated, now takes the form

$$\begin{aligned} ln\mathcal{Z} &= N ln z \\ &\cong N\left\{ln\left[V(2\pi M k_B T)^{\frac{3}{2}}\right] + ln z_{\text{rot}} + ln z_{\text{vib}}\right\} \end{aligned} \qquad (16.21b)$$

It receives independent additive contributions from the translational motion of the center-of-mass of the molecule, from the rotation of the (rigid) dumbbell about the center-of-mass, and from the remaining vibration of the radial coordinate about its equilibrium value r_0.

In accord with the qualitative illustration in Figure 16.2b, the radial vibrations are assumed to change the radial coordinate by only a very small amount, so that the rotation of the system can be analyzed as if the molecule were rigid with a well-defined radius r_0. In the present framework, the difference between the hamiltonians of Eq.(16.7b) and (16.17b) results in an additional interaction term describing rotation-vibration coupling, whose effects must be included in higher order.

Let us then proceed to analyze the partition function of the rigid rotor

$$z_{\text{rot}} = \int dp_\theta dp_\phi d\theta d\phi \exp\left\{-\beta\left[\frac{1}{2\mu r_0^2}\left(p_\theta^2 + \frac{p_\phi^2}{\sin^2\theta}\right)\right]\right\} \qquad (16.22b)$$

With the introduction of the following variables

$$x_1 \equiv \left(\frac{\beta}{2\mu}\right)^{\frac{1}{2}}\frac{p_\theta}{r_0} \qquad (16.23b)$$

$$x_2 \equiv \left(\frac{\beta}{2\mu}\right)^{\frac{1}{2}}\frac{p_\phi}{r_0 \sin\theta}$$

the momentum differential takes the form

$$dp_\theta dp_\phi = \frac{2\mu r_0^2}{\beta}\sin\theta \ dx_1 dx_2 \qquad (16.24b)$$

and the partition function becomes

$$z_{\text{rot}} = \frac{2\mu r_0^2}{\beta} \int_{-\infty}^{\infty} e^{-x_1^2} dx_1 \int_{-\infty}^{\infty} e^{-x_2^2} dx_2 \int_0^{\pi} \sin\theta d\theta \int_0^{2\pi} d\phi \quad (16.25b)$$

The gaussian integrals are evaluated in Eq.(13.13), and the remaining angular integrals just give the total solid angle 4π

$$z_{\text{rot}} = \frac{8\pi^2 \mu r_0^2}{\beta} = 8\pi^2 \mu r_0^2 k_B T \quad (16.26b)$$

With the retention of the translation and rotation contributions in Eq.(16.21b), the logarithm of the partition function for a collection of rigid rotors becomes

$$ln\mathcal{Z} = N\left\{ ln\left[V\left(\frac{2\pi M}{\beta}\right)^{\frac{3}{2}} \right] + ln\left[\frac{8\pi^2 \mu r_0^2}{\beta} \right] \right\}$$

$$= N\left\{ lnV - \frac{5}{2}ln\beta + \text{const.} \right\} \qquad \text{; rigid rotors} \quad (16.27b)$$

The pressure of the system follows from Eqs.(13.29-30), and the energy and heat capacity from Eqs.(13.21-23)

$$p = \frac{1}{\beta}\frac{\partial ln\mathcal{Z}}{\partial V} = \frac{N}{\beta V} = \frac{Nk_B T}{V} \quad (16.28b)$$

$$E = -\frac{\partial ln\mathcal{Z}}{\partial \beta} = \frac{5}{2}\frac{N}{\beta} = \frac{5}{2}Nk_B T \quad (16.29b)$$

$$C_V = \frac{\partial E}{\partial T} = \frac{5}{2}Nk_B \qquad \text{; rigid rotors} \quad (16.30b)$$

Note that relative to the ideal monatomic gas [Eqs.(13.22-23)], an ideal gas of rigid rotors has an extra contribution of $Nk_B T$ in the internal energy and Nk_B in the heat capacity. The equipartition theorem provides an immediate explanation of these results, since the additional rotational hamiltonian appearing in Eqs.(16.17b) and (16.19b) is a positive homogeneous quadratic form ($\nu = 2$) in the two angular momentum variables (p_θ, p_ϕ) of each molecule ($g = 2N$); thus there is an additional contribution to the energy of the system of $E_{\text{rot}} = (g/\nu)k_B T = Nk_B T$.

Vibrations. We now consider the additional contribution of the vibrations in the radial coordinate of the diatomic molecule. The internal vibrational partition function contributes additively in Eq.(16.21b); it is given by Eq.(16.20b)

$$z_{\text{vib}} = \int_{-\infty}^{\infty} e^{-\beta p_r^2/2\mu} \, dp_r \int_{0}^{\infty} e^{-\beta U(r)} dr \tag{16.31b}$$

The following change of variable

$$x = \left(\frac{\beta}{2\mu}\right)^{\frac{1}{2}} p_r$$

$$dx = \left(\frac{\beta}{2\mu}\right)^{\frac{1}{2}} dp_r \tag{16.32b}$$

permits the integral over the radial momenta to be immediately performed as above

$$\int_{-\infty}^{\infty} e^{-\beta p_r^2/2\mu} dp_r = \left(\frac{2\mu}{\beta}\right)^{\frac{1}{2}} \int_{-\infty}^{\infty} e^{-x^2} dx = \left(\frac{2\pi\mu}{\beta}\right)^{\frac{1}{2}} \tag{16.33b}$$

In evaluating the remaining integral over the radial coordinate, it is convenient to first simplify the expression by making use of the fact that the zero of energy is arbitrary [see the discussion following Eq.(10.47)]. We are therefore free to define

$$U(r_0) \equiv 0 \tag{16.34b}$$

Any displacement of the radial coordinate from its equilibrium value $r = r_0$ now raises the energy of the system relative to that of the previously discussed rigid rotors, as indicated in Figure 16.3b. It takes a positive amount of energy, the binding energy (B.E.), to dissociate the molecule, and this is also indicated in Figure 16.3b.

Let us define the deviation of the radial coordinate r from its equilibrium value as

$$\delta \equiv r - r_0 \tag{16.35b}$$

We now observe that the exponential factor $\exp[-U(r)/k_BT]$ vanishes whenever $U(r)/k_BT \gg 0$. It is evident from Figure 16.3b that if the minimum in the interatomic potential is deep and sharp enough, and $k_BT \ll (B.E.)$, then only a very small region in δ will contribute to the vibrational partition function. For orientation, the binding energy of most molecules at room temperature is $(B.E.) \cong 1eV$ while $k_BT \cong (1/40)eV$.

Let us assume that these conditions are satisfied, so that the situation illustrated in Figure 16.3b applies and

$$k_B T/(B.E.) \ll 1$$
$$|\delta|/r_0 \ll 1 \qquad (16.36b)$$

Then over a small region in the vicinity of the minimum, the curve in Figure 16.3b can always be *approximated by a parabola*. This follows from a Taylor series expansion of the potential

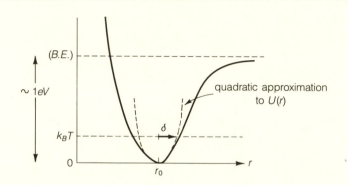

Figure 16.3b Interatomic potential $U(r)$ appearing in the internal vibrational partition function. The zero of energy is arbitrary and is here defined by $U(r_0) \equiv 0$. It takes a positive amount of energy, the binding energy (B.E.), to dissociate the molecule. The situation $k_B T \ll (B.E.)$ is illustrated. The deviation of the radial coordinate from its equilibrium value is denoted by $\delta = r - r_0$, and the quadratic approximation to $U(r)$ obtained by keeping the leading terms in the Taylor series expansion about r_0 in Eq. (16.37b) is also indicated.

$$
\begin{aligned}
U(r) &= U(r_0 + \delta) \\
&\cong U(r_0) + \delta \left(\frac{\partial U}{\partial r} \right)_{r_0} + \frac{1}{2}\delta^2 \left(\frac{\partial^2 U}{\partial r^2} \right)_{r_0} \qquad (16.37b)
\end{aligned}
$$

The first two terms in this expansion vanish by Eq.(16.34b) and the condition that r_0 is a minimum of the potential

$$\left(\frac{\partial U}{\partial r} \right)_{r_0} = 0 \qquad (16.38b)$$

It is evident from Figure 16.3b that the remaining term in this expansion will be positive definite.

With the substitution of the quadratic approximation to the true potential in Eq.(16.37b), the radial integral in the vibrational partition function becomes

$$\int_0^\infty e^{-\beta U(r)}\,dr \cong \int_{-\infty}^\infty \exp\left[\frac{-\beta}{2}\left(\frac{\partial^2 U}{\partial r^2}\right)_{r_0}\delta^2\right]d\delta \tag{16.39b}$$

Here the integration variable has been changed to δ using Eq.(16.35b), and the exponential damping in the integrand permits the limits of integration on δ to be extended to $\pm\infty$ with negligible error. It is convenient to define the new quantities

$$\left(\frac{\partial^2 U}{\partial r^2}\right)_{r_0} \equiv k \equiv \mu\omega^2 \tag{16.40b}$$

where k is an equivalent "spring constant" and ω an "oscillator frequency." The integral in Eq.(16.39b) becomes

$$\int_0^\infty e^{-\beta U(r)}\,dr \cong \int_{-\infty}^\infty e^{-\beta\mu\omega^2\delta^2/2}\,d\delta \tag{16.41b}$$

A change of variable

$$x \equiv \left(\frac{\beta\mu}{2}\right)^{\frac{1}{2}}\omega\delta$$

$$dx = \left(\frac{\beta\mu}{2}\right)^{\frac{1}{2}}\omega\,d\delta \tag{16.42b}$$

converts it to a familiar integral

$$\int_0^\infty e^{-\beta U(r)}\,dr \cong \left(\frac{2}{\beta\mu\omega^2}\right)^{\frac{1}{2}}\int_{-\infty}^\infty e^{-x^2}\,dx = \left(\frac{2\pi}{\beta\mu\omega^2}\right)^{\frac{1}{2}} \tag{16.43b}$$

The internal vibrational partition function in Eq.(16.31b) is now obtained from Eqs.(16.33b) and (16.43b); it takes the form

$$\begin{aligned}z_{\text{vib}} &\cong \left(\frac{2\pi\mu}{\beta}\right)^{\frac{1}{2}}\left(\frac{2\pi}{\beta\mu\omega^2}\right)^{\frac{1}{2}} = \frac{2\pi}{\beta\omega}\\ &= \frac{2\pi}{\omega}k_B T\end{aligned} \tag{16.44b}$$

The partition function for an ideal gas of diatomic molecules under the assumptions of Eqs.(16.17b) and (16.36b) is obtained by combining Eqs.(16.13b), (16.15b), (16.18b), (16.26b), and (16.44b)

$$\mathcal{Z} = z^N$$

$$z = \left[V \left(\frac{2\pi M}{\beta} \right)^{\frac{3}{2}} \right] \cdot \left[\frac{8\pi^2 \mu r_0^2}{\beta} \right] \cdot \left[\frac{2\pi}{\beta \omega} \right] \quad ; \text{ diatomic gas}$$

$$(16.45b)$$

The logarithm of this expression follows as in Eq.(16.21b)

$$ln\mathcal{Z} = N \left\{ lnV - \frac{7}{2} ln\beta + \text{const.} \right\} \quad ; \text{ diatomic gas} \qquad (16.46b)$$

The pressure, energy, and heat capacity are determined as in Eqs.(16.28b-30b)

$$p = \frac{1}{\beta} \frac{\partial ln\mathcal{Z}}{\partial V} = \frac{N}{\beta V} = \frac{Nk_B T}{V} \qquad (16.47b)$$

$$E = -\frac{\partial ln\mathcal{Z}}{\partial \beta} = \frac{7}{2} \frac{N}{\beta} = \frac{7}{2} Nk_B T \qquad (16.48b)$$

$$C_V = \frac{\partial E}{\partial T} = \frac{7}{2} Nk_B \qquad ; \text{ diatomic gas} \qquad (16.49b)$$

As with the system of rigid rotors, the ideal gas law again holds. This arises from the center-of-mass motion of the molecules that gives rise to precisely the same pressure as obtained for the ideal monatomic gas (Sections 13 and 14); no interactions *between* the molecules have been included.   In contrast to the system of rigid rotors described by Eqs.(16.29b-16.30b), the energy is greater by $Nk_B T$ and the heat capacity by Nk_B when the internal radial vibrations of the diatomic molecule are included.   The equipartition theorem again provides an understanding of this result.   The additional effective hamiltonian appearing in the vibrational partition function in Eqs.(16.33b) and (16.39b) is a homogeneous quadratic form ($\nu = 2$) in the dynamical variables (p_r, δ) for each of the N molecules ($g = 2N$). Thus there is an additional contribution to the energy of $E_{\text{vib}} = (g/\nu)k_B T = Nk_B T$ coming from the radial vibrations.

Let us reflect on the results we have obtained.   Expressions for the constant-volume heat capacity of ν_m moles of various distinct physical systems have been derived. If the molar heat capacity is denoted by

$$c_v \equiv \frac{C_V}{\nu_m} \qquad (16.50b)$$

then

$$c_v/R = 3/2 \; ; \; \text{ideal monatomic gas}$$
$$= 5/2 \; ; \; \text{ideal gas of rigid diatomic molecules}$$
$$= 7/2 \; ; \; \text{ideal gas of diatomic molecules including vibrations}$$
$$(16.51b)$$

Here R is the gas constant of Eq.(10.43). But what is going on? If one makes the oscillator frequency in Eq.(16.40b) larger and larger, then it is clear from Figure 16.3b that the potential becomes stiffer and stiffer (the curvature at the minimum increases), and the molecule becomes more and more rigid. In contrast to the result $c_v = 5R/2$ obtained for rigid rotors, however, one finds $c_v = 7R/2$ when vibrations are included *independent of the value of ω!* Thus we are faced with an apparent contradiction; we appear to get two distinct answers in the limit $\omega \longrightarrow \infty$.

The resolution to this dilemma comes from *quantum mechanics* and *quantum statistics*. We shall see that the result $c_v = 5R/2$ holds at low temperatures as long as $\hbar\omega >> k_B T$, where $\hbar = h/2\pi$ and h is Planck's constant

$$\hbar \equiv \frac{h}{2\pi} = 1.055 \times 10^{-27} \text{erg sec} \qquad ; \text{h is Planck's constant} \qquad (16.52b)$$

At high temperatures where $\hbar\omega << k_B T$, one must use the result $c_v = 7R/2$ for the diatomic molecule. Thus

$$
\begin{aligned}
c_v &= \frac{5}{2}R & k_B T << \hbar\omega \\
&= \frac{7}{2}R & k_B T >> \hbar\omega
\end{aligned}
\qquad (16.53b)
$$

In contrast to the classical results that hold at all temperatures, quantum statistics implies that the vibrational degrees of freedom are "frozen out" as the temperature is lowered.

There is a similar relation concerning the rotational degrees of freedom of the diatomic molecule in quantum statistics. The result $c_v = 5R/2$ holds as long as $\hbar^2/\mu r_0^2 << k_B T$. At low temperature where $\hbar^2/\mu r_0^2 >> k_B T$, the heat capacity reduces to the result $c_v = 3R/2$ for the monatomic gas

$$
\begin{aligned}
c_v &= \frac{3}{2}R & k_B T << \frac{\hbar^2}{\mu r_0^2} \\
&= \frac{5}{2}R & k_B T >> \frac{\hbar^2}{\mu r_0^2}
\end{aligned}
\qquad (16.54b)
$$

The rotational degrees of freedom are frozen out as well, as the temperature is lowered. Since it is generally true for the diatomic molecule that

$$\frac{\hbar^2}{\mu r_0^2} << \hbar\omega \qquad (16.55b)$$

the vibrational degrees of freedom will be frozen out first, and then the rotational degrees of freedom, as the temperature is lowered. The experimentally observed heat capacity of a typical diatomic gas is sketched in Figure 16.4b.

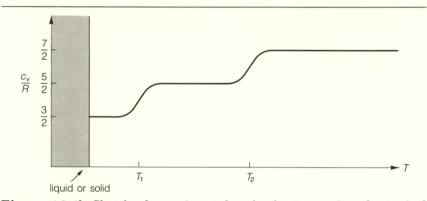

Figure 16.4b Sketch of experimental molar heat capacity of a typical diatomic gas. Here $k_B T_2 = \hbar\omega$ is the characteristic temperature below which the vibrational degrees of freedom are frozen out, and $k_B T_1 \equiv \hbar/\mu r_0^2$ is the corresponding temperature for the rotational degrees of freedom. It is assumed in this sketch that $k_B T << (B.E.)$, so that the molecule does not dissociate over this temperature interval. Dissociation will always occur at high enough temperature. Furthermore, at very low temperatures one will have either a liquid or a solid.

Thus we have our first example of a paradox in classical statistical mechanics that is resolved by quantum mechanics and quantum statistics.

c) Solids. Consider N atoms of mass m bound in a solid. It shall be assumed that the solid is a crystal and that each atom has a stable equilibrium position. Furthermore, the atoms on the surface of the crystal will be held fixed to prevent translations and rotations of the body. The configuration of the solid is described by specifying the coordinate vector $\mathbf{r}_s = (x_s, y_s, z_s)$ of each atom measured from a stationary origin O (Figure

16.1c). The corresponding momentum will be denoted by $\mathbf{p}_s$. The index $s = 1, 2, \ldots, N$ runs over the atoms.

The hamiltonian for this system has the general form

$$H = \sum_{s=1}^{N} \frac{\mathbf{p}_s^2}{2m} + U(\mathbf{r}_1, \mathbf{r}_2 \ldots, \mathbf{r}_N) \qquad (16.1c)$$

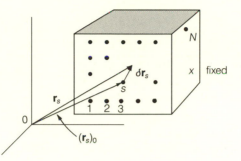

Figure 16.1c Coordinate vectors describing the configuration of a solid. Here $(\mathbf{r}_s)_0$ denotes the equilibrium position of the sth *atom*.

Let $(\mathbf{r}_s)_0$ denote the equilibrium position of the sth atom. Since equilibrium is reached at the absolute minimum of the potential U, the equilibrium positions are determined by the conditions

$$\frac{\partial U}{\partial x_s} = 0 \qquad ; \quad \frac{\partial U}{\partial y_s} = 0 \qquad ; \quad \frac{\partial U}{\partial z_s} = 0 \qquad (16.2c)$$

Now let

$$\mathbf{r}_s = (\mathbf{r}_s)_0 + \delta\mathbf{r}_s \qquad (16.3c)$$

where $\delta\mathbf{r}_s$ is the displacement of the sth atom from its equilibrium position (Figure 16.1c). The potential in Eq.(16.1c) then takes the form

$$U = U\left[(\mathbf{r}_1)_0 + \delta\mathbf{r}_1, \ldots, (\mathbf{r}_s)_0 + \delta\mathbf{r}_s, \ldots, (\mathbf{r}_N)_0 + \delta\mathbf{r}_N\right] \qquad (16.4c)$$

As long as the temperature is well below the melting point of the solid, one may assume that $\delta\mathbf{r}_s$ is sufficiently small to allow a Taylor series expansion of U up to quadratic terms in the components of $\delta\mathbf{r}_s$. The

Cartesian components of $\delta\mathbf{r}_s$ and of the momenta $\mathbf{p}_s$ will now be used as the generalized coordinates and momenta for the system. They will be ordered and labeled with an index i $\{p_i, q_i : i = 1, 2, \ldots, 3N\}$, as in Section 1

$$
\begin{aligned}
p_{1x} &= p_1, & p_{1y} &= p_2, & p_{1z} &= p_3, & \ldots, & p_{Nz} &= p_{3N} \\
\delta r_{1x} &= q_1, & \delta r_{1y} &= q_2, & \delta r_{1z} &= q_3, & \ldots, & \delta r_{Nz} &= q_{3N}
\end{aligned}
\tag{16.5c}
$$

With $(\mathbf{r}_s)_0$ determined by equilibrium, the hamiltonian takes the form

$$
H = \sum_{i=1}^{3N} \frac{p_i^2}{2m} + U(q_1, q_2 \ldots, q_i, \ldots, q_{3N})
\tag{16.6c}
$$

Expansion of the potential for small q_i gives

$$
U \cong U_0 + \sum_{i=1}^{3N} \left(\frac{\partial U}{\partial q_i} \right)_0 q_i + \frac{1}{2} \sum_{i,j=1}^{3N} \left(\frac{\partial^2 U}{\partial q_i \partial q_j} \right)_0 q_i q_j
\tag{16.7c}
$$

Here U_0 is the value of U at the absolute minimum achieved at $q_i = 0$ and

$$
\left(\frac{\partial U}{\partial q_i} \right)_0 = 0
\tag{16.8c}
$$

Since the energy scale is arbitrary, one is again free to define

$$
U_0 \equiv 0
\tag{16.9c}
$$

The second derivatives of the potential evaluated at equilibrium will be defined as

$$
\left(\frac{\partial^2 U}{\partial q_i \partial q_j} \right)_0 \equiv a_{ij} = a_{ji}
\tag{16.10c}
$$

With the neglect of terms that are of higher order in q_i, one thus finds

$$
U \cong \frac{1}{2} \sum_{i,j=1}^{3N} a_{ij} q_i q_j
\tag{16.11c}
$$

This expression must be *positive definite* if the potential indeed has an absolute minimum $U_0 = 0$ for $q_i = 0$. Within this framework, the hamiltonian takes the form

$$
H = \sum_{i=1}^{3N} \frac{p_i^2}{2m} + \frac{1}{2} \sum_{i,j=1}^{3N} a_{ij} q_i q_j
\tag{16.12c}
$$

This expression is a positive definite quadratic form in the $6N$ variables $\{p_i, q_i\}$. The equipartition theorem of Eq.(15.12) with $\nu = 2$ and $g = 6N$ then yields the following expression for the total energy of the system

$$E = \frac{6Nk_BT}{2} = 3Nk_BT \qquad (16.13c)$$

The constant-volume heat capacity is given by

$$C_V = \left(\frac{\partial E}{\partial T}\right)_V = 3Nk_B \qquad (16.14c)$$

This expression holds for ν_m moles, and the molar heat capacity $c_v = C_V/\nu_m$ is given by

$$c_v = 3R \qquad (16.15c)$$

where R is the gas constant of Eq.(10.43). This is the celebrated "Law of Dulong and Petit."

The relation (16.15c) is observed to hold experimentally for solids over a range of temperature well below the melting point *but not too low*. In fact, the *Nernst theorem* demands generally that

$$c_v \longrightarrow 0 \quad for \quad T \longrightarrow 0 \qquad ; \; Nernst \; Theorem \qquad (16.16c)$$

The observed specific heat of a typical solid is sketched in Figure 16.2c.

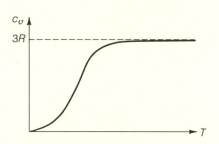

Figure 16.2c. Sketch of observed molar specific heat of a typical solid.

The explanation of the low-temperature behavior is again found in quantum theory and involves the "freezing out" of oscillations if $k_BT \ll$

$\hbar\omega$. To proceed, it is necessary to know something about the actual frequency spectrum of oscillations in a crystal.

Normal Coordinates. The introduction of normal coordinates allows the hamiltonian in Eq.(16.12c) to be written as the sum of squares, rather than just as a quadratic form; it then separates into a sum of hamiltonians, each of which describes a simple harmonic oscillator.

Instead of the coordinates q_i $(i = 1, 2, \ldots, n)$ where $n = 3N$, introduce a new set of coordinates q'_k $(k = 1, 2, \ldots, n)$ such that

$$q_i = \sum_{k=1}^{n} c_{ik} q'_k \qquad ; \; i = 1, 2, \ldots, n \qquad (16.17c)$$

Then

$$q_j = \sum_{l=1}^{n} c_{jl} q'_l \qquad (16.18c)$$

and the potential term in the hamiltonian of Eq.(16.12c) takes the form

$$U = \frac{1}{2} \sum_{i,j=1}^{n} a_{ij} q_i q_j = \frac{1}{2} \sum_{k,l=1}^{n} \left(\sum_{i,j=1}^{n} c_{ik} c_{jl} a_{ij} \right) q'_k q'_l \qquad (16.19c)$$

The trick is to determine the coefficients c_{ik} so that the quantity in parentheses in the last term only contributes when $k = l$, whereupon it takes the value λ_k. Thus we seek coefficients c_{ik} such that

$$\sum_{i,j=1}^{n} c_{ik} c_{jl} a_{ij} = \lambda_k \delta_{kl} \qquad (16.20c)$$

where the Kronecker delta function is defined by

$$\begin{aligned} \delta_{kl} &= 1 \qquad if \; k = l \\ &= 0 \qquad if \; k \neq l \end{aligned} \qquad (16.21c)$$

In this case, the potential reduces to a sum of squares in terms of the new normal coordinates

$$U = \frac{1}{2} \sum_{k=1}^{n} \lambda_k q'^2_k \qquad (16.22c)$$

In order to determine the appropriate coefficients c_{ik}, consider the following set of linear equations

$$\sum_{j=1}^{n} a_{ij} c_j = \lambda c_i \qquad ; \; i = 1, 2 \ldots, n \qquad (16.23c)$$

The a_{ij} defined in Eq.(16.10c) form a real, symmetric matrix of constant coefficients determined by properties of the potential, and this set of linear, homogeneous, algebraic equations for the quantities $\{c_i;\ i = 1, 2, \ldots, n\}$ will only possess a non-trivial solution if the determinant of the coefficients vanishes

$$\det [a_{ij} - \lambda \delta_{ij}] = 0 \qquad (16.24c)$$

The evaluation of the $n \times n$ determinant yields an nth-order polynomial in the quantity λ. This polynomial will have precisely n roots; we refer to these roots as the *eigenvalues* and label them as follows

$$\lambda_s \qquad ;\ s = 1, 2, \ldots, n \qquad (16.25c)$$

The corresponding solutions to the linear Eqs.(16.23c), the *eigenvectors*, will be denoted by

$$c_i^{(s)} \qquad ;\ i = 1, 2, \ldots, n \qquad (16.26c)$$

They satisfy the equations

$$\sum_{j=1}^{n} a_{ij} c_j^{(s)} = \lambda_s c_i^{(s)} \qquad ;\ i = 1, 2, \ldots, n \qquad (16.27c)$$

We proceed to show that these coefficients will do the job for us and will indeed lead to Eq.(16.20c). Write Eqs.(16.27c) for two different eigenvalues and corresponding eigenvectors

$$\sum_{j=1}^{n} a_{ij} c_j^{(k)} = \lambda_k c_i^{(k)} \qquad ;\ i = 1, 2, \ldots, n$$
$$\sum_{j=1}^{n} a_{ij} c_j^{(l)} = \lambda_l c_i^{(l)} \qquad (16.28c)$$

Multiply the first set of equations by $c_i^{(l)}$ and sum on i. Similarly, multiply the second set of equations by $c_i^{(k)}$ and sum on i. Now subtract the two expressions. The lefthand side of the result gives

$$\sum_{i,j=1}^{n} \left(a_{ij} c_i^{(l)} c_j^{(k)} - a_{ij} c_i^{(k)} c_j^{(l)} \right) = 0 \qquad (16.29c)$$

A change of dummy summation indices $i \leftrightarrow j$ in the second term, and the observation that $a_{ij} = a_{ji}$ is symmetric, shows that this expression vanishes as indicated. The righthand side of the result must therefore also vanish

$$(\lambda_k - \lambda_l) \sum_{i=1}^{n} c_i^{(l)} c_i^{(k)} = 0 \tag{16.30c}$$

If the eigenvalues are distinct $\lambda_k \neq \lambda_l$, one concludes that

$$\sum_{i=1}^{n} c_i^{(l)} c_i^{(k)} = 0 \qquad ; \; \lambda_k \neq \lambda_l \tag{16.31c}$$

The eigenvectors corresponding to distinct eigenvalues are therefore *orthogonal*. Since Eqs.(16.23c) are homogeneous, the coefficients $c_i^{(k)}$ can be arbitrarily normalized with an overall scale factor. We choose the normalization condition

$$\sum_{i=1}^{n} c_i^{(l)} c_i^{(k)} = \delta_{lk} \tag{16.32c}$$

If it should turn out that the eigenvalues are degenerate $\lambda_k = \lambda_l$, the corresponding eigenvectors can still be orthogonalized by means of the Schmidt procedure (Problem 16.c3), and the orthonormality relation in Eq.(16.32c) can be assumed to hold quite generally.

Now use the coefficients determined through the above procedure to carry out the transformation to new coordinates in Eq.(16.17c). Define

$$c_{ik} \equiv c_i^{(k)} \qquad ; \; i, k = 1, 2, \ldots, n \tag{16.33c}$$

Evaluation of the bilinear form on the lefthand side of Eq.(16.20c) then yields the expression

$$\sum_{i,j=1}^{n} a_{ij} c_{ik} c_{jl} = \sum_{i=1}^{n} c_i^{(k)} \left(\sum_{j=1}^{n} a_{ij} c_j^{(l)} \right) = \sum_{i=1}^{n} c_i^{(k)} \left(\lambda_l c_i^{(l)} \right)$$
$$= \lambda_l \sum_{i=1}^{n} c_i^{(k)} c_i^{(l)} \tag{16.34c}$$

Here the eigenvalue Eqs.(16.27c) have been introduced. The orthonormality condition of Eq.(16.32c) then leads to the desired result

$$\sum_{i,j=1}^{n} a_{ij} c_{ik} c_{jl} = \lambda_l \delta_{kl} \tag{16.35c}$$

The eigenvalues determined by this procedure will be real. This is immediately established by multiplying the defining Eqs.(16.23c) by c_i^*, summing on i, and solving for λ

$$\lambda = \frac{\sum_{i,j=1}^{n} c_i^* a_{ij} c_j}{\sum_{i=1}^{n} c_i^* c_i} \tag{16.36c}$$

Now take the complex conjugate of this expression. Since a_{ij} is a real symmetric matrix, a change of dummy summation indices establishes that

$$\lambda^* = \lambda \quad ; \text{ real} \tag{16.37c}$$

is real as claimed.

When the determinant of coefficients vanishes, one of the linear Eqs. (16.23c), say the nth, becomes linearly dependent and may be discarded. Divide the remaining Eqs.(16.27c) by $c_n^{(s)}$ (assumed non-zero). One is then left with $n - 1$ linear, *inhomogeneous* equations for the $n - 1$ ratios $c_i^{(s)}/c_n^{(s)}$; $i = 1, 2, \ldots, n - 1$. These inhomogeneous equations yield a *unique solution* for the $n - 1$ ratios $c_i^{(s)}/c_n^{(s)}$ and the normalization condition of Eq.(16.32c) then determines the coefficients $c_i^{(s)}$; $i = 1, 2, \ldots, n$ completely.† Since the linear equations and normalization condition are real, *both* the eigenvalues and eigenvectors in Eqs.(16.25c) and (16.26c) generated through this procedure will be *real*.

Since the potential was assumed to be at an absolute minimum, the result in Eq.(16.22c) must be a positive definite quadratic form. We conclude that the eigenvalues determined by the above analysis must therefore also be *positive*.

Consider now the transformation to normal coordinates in Eq.(16.17c) to be a point transformation as defined in Eq.(1.47). The normal coordinates are defined by the relations

$$q_i = \sum_{k=1}^{n} c_{ik} q_k' \quad ; i = 1, 2, \ldots, n \tag{16.38c}$$

where the transformation coefficents are the eigenvectors in Eq.(16.33c); they satisfy the orthonormality relation of Eq.(16.32c)

$$\sum_{i=1}^{n} c_{ik} c_{il} = \delta_{kl} \tag{16.39c}$$

---

† Up to an overall sign.

This relation can be used to invert Eqs.(16.38c)

$$q'_k = \sum_{i=1}^{n} c_{ik} q_i \qquad ; \; k = 1, 2, \ldots, n \qquad (16.40c)$$

to obtain a result which is explicitly in the form of Eqs.(1.47). Differentiation of Eqs.(16.38c) and (16.40c) leads to the relations

$$\frac{\partial q_i}{\partial q'_k} = c_{ik} = \frac{\partial q'_k}{\partial q_i} \qquad (16.41c)$$

where the partial derivative indicates that the other members of the appropriate variable sets in these equations are to be held fixed.

New momenta may be correspondingly defined by the relations

$$p_i = \sum_{k=1}^{n} c_{ik} p'_k \qquad ; \; i = 1, 2, \ldots, n \qquad (16.42c)$$

They can be similarly inverted to give

$$p'_k = \sum_{i=1}^{n} c_{ik} p_i \qquad ; \; k = 1, 2, \ldots, n \qquad (16.43c)$$

Substitution of Eqs.(16.41c) leads to the results

$$p'_k = \sum_{i=1}^{n} \frac{\partial q_i}{\partial q'_k} p_i \qquad (16.44c)$$

and

$$p_i = \sum_{k=1}^{n} \frac{\partial q'_k}{\partial q_i} p'_k \qquad (16.45c)$$

Since the first of these relations is identical to Eq.(1.49), the point transformation defined by Eqs.(16.40c) and (16.43c) is a *canonical transformation*. It is evident from the orthogonality relation in Eq.(16.39c) that the length of the vector **p** is preserved by this transformation

$$\sum_{i=1}^{n} p_i p_i = \sum_{k=1}^{n} p'_k p'_k \qquad (16.46c)$$

Hence the kinetic energy can be written in terms of the new momenta according to

$$T = \sum_{i=1}^{n} \frac{p_i^2}{2m} = \sum_{k=1}^{n} \frac{p_k'^2}{2m} \qquad (16.47c)$$

Note that it is assumed here that all the atoms in the solid have the same mass. With the aid of Eqs.(16.22c) and (16.47c), the hamiltonian in Eq.(16.12c) can be written as

$$H = \sum_{k=1}^{n} \left[\frac{p_k'^2}{2m} + \frac{1}{2}\lambda_k q_k'^2 \right] \qquad (16.48c)$$

It is written in terms of the normal coordinates and momenta $\{p_k', q_k'; k = 1, 2, \ldots, n\}$. One is now free to take these quantities as new generalized coordinates and momenta for the problem. Hamilton's equations in terms of these new variables then lead to

$$\dot{q}_k' = \frac{\partial H}{\partial p_k'} = \frac{p_k'}{m} \qquad ; \ k = 1, 2, \ldots, n$$

$$-\dot{p}_k' = \frac{\partial H}{\partial q_k'} = \lambda_k q_k' \qquad (16.49c)$$

A combination of these results gives

$$\ddot{q}_k' = -\omega_k^2 q_k' \qquad ; \ k = 1, 2, \ldots, n \qquad (16.50c)$$

where the oscillator frequencies are defined by

$$\lambda_k \equiv m\omega_k^2 \qquad (16.51c)$$

The hamiltonian in Eq.(16.48c) describes the small oscillations about equilibrium of the atoms in a crystal. When written in terms of the normal coordinates and momenta, obtained from the original coordinates and momenta in Eq.(16.5c) by the constructed canonical transformation, the hamiltonian is *diagonalized* and describes a system of *uncoupled simple harmonic oscillators* each of which oscillates with one of the *normal mode frequencies* determined by the eigenvalues in Eq.(16.25c). The solution to Eqs.(16.50c) can be written immediately as†

$$q_k' = A_k cos(\omega_k t + \phi_k) \qquad ; \ k = 1, \ldots, n$$

$$p_k' = m\dot{q}_k' = -m\omega_k A_k sin(\omega_k t + \phi_k) \qquad (16.52c)$$

---

† We here and henceforth in sections 16c and 16d suppress the explicit indication of a time dependence in the notation used on the lefthand side of these equations. It is understood that all the dynamic variables carry a time dependence governed by Hamilton's equations.

The constants $\{A_k, \phi_k; \; i = 1, 2, \ldots, n\}$ are determined by the intitial conditions.

If we continue the discussion following Eq.(16.16c), then to an excellent approximation, all modes with frequencies $\hbar\omega >> k_B T$ are unexcited at the temperature T; one can say that these modes are "frozen out." The problem then is to determine the spectrum of normal mode frequencies for a crystal. To this end one can consider various simple models of a crystal, and we first examine the *linear chain* for which the normal mode spectrum can be calculated relatively simply.

Linear Chain. Consider a linear chain of identical atoms separated in their equilibrium configuration by a common distance a (Figure 16.3c). We assume the atoms are constrained to move only *longitudinally* along the x axis. Let $(x_s)_0$ denote the equilibrium coordinate of the sth particle, and δx_s denote its displacement from equilibrium. The hamiltonian for this system takes the form

$$H = \sum_{s=1}^{N} \frac{p_s^2}{2m} + U \qquad (16.53c)$$

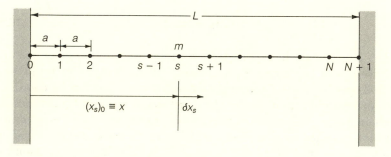

Figure 16.3c. The linear chain with fixed endpoints.

A typical interatomic potential is sketched in Figure 16.4c as a function of the relative separation r of the two atoms. It will be assumed that $k_B T$ is low enough so that the displacement of the atom from its equilibrium position δx_s is small. It will further be assumed that $U \approx 0$ at $r = 2a$ so that only *nearest neighbor interactions* need be taken into account in the potential energy, which thus takes the form

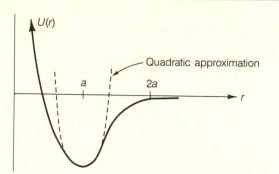

Figure 16.4c A typical interatomic potential in the linear chain as a function of the relative separation r.

$$U = U_{01} + U_{12} + U_{23} + \ldots + U_{N,N+1} = \sum_{s=0}^{N} U_{s,s+1} \qquad (16.54c)$$

where

$$U_{s,s+1} = U\left(\left[(x_{s+1})_0 + \delta x_{s+1}\right] - \left[(x_s)_0 + \delta x_s\right]\right) = U\left(a + \delta x_{s+1} - \delta x_s\right) \qquad (16.55c)$$

The boundary condition of fixed endpoints will be assumed so that

$$\delta x_0 = \delta x_{N+1} = 0 \qquad ; \text{ fixed endpoints} \qquad (16.56c)$$

The displacements from equilibrium will now be taken as the generalized coordinates

$$q_s = \delta x_s \qquad ; s = 1, 2, \ldots, N \qquad (16.57c)$$

so that the hamiltonian for the linear chain takes the form

$$H = \sum_{s=1}^{N} \frac{p_s^2}{2m} + U$$

$$U = \sum_{s=0}^{N} U\left(a + q_{s+1} - q_s\right) \qquad (16.58c)$$

and the condition of fixed endpoints is

$$q_0 = q_{N+1} = 0 \qquad ; \text{ fixed endpoints} \qquad (16.59c)$$

If $k_B T$ is such that the displacements are indeed small, then the quadratic approximation to the interatomic potential in the vicinity of its minimum will suffice, and it will be written as

$$U \cong U_0 + \sum_{s=1}^{N} \left(\frac{\partial U}{\partial q_s} \right)_0 q_s + \frac{1}{2} \sum_{s=1}^{N} \sum_{s'=1}^{N} \left(\frac{\partial^2 U}{\partial q_s \partial q_{s'}} \right)_0 q_s q_{s'} \qquad (16.60c)$$

The constant term U_0 is the equilibrium binding energy between atoms. Since the energy scale in the problem is again arbitrary, we define $U_0 \equiv 0$. By the definition of the minimum, the first derivatives of the potential must also vanish there. Thus one has

$$U_0 = 0$$
$$\left(\frac{\partial U}{\partial q_s} \right)_0 = 0 \quad ; \ s = 1, 2, \ldots, N \qquad (16.61c)$$

The remaining term in the potential in Eq.(16.60c) is defined as

$$a_{ss'} \equiv \left(\frac{\partial^2 U}{\partial q_s \partial q_{s'}} \right)_0 \quad ; \ s, s' = 1, 2, \ldots, N \qquad (16.62c)$$

where

$$a_{ss} = \left(\frac{\partial^2 U}{\partial q_s^2} \right)_0 \qquad (16.63c)$$

The appropriate derivatives of the potential in Eq.(16.58c) can now be evaluated. Since there are two terms that contain the coordinate q_s

$$a_{ss} = \frac{\partial^2 U}{\partial q_s^2}(a + q_{s+1} - q_s)\bigg|_0 + \frac{\partial^2 U}{\partial q_s^2}(a + q_s - q_{s-1})\bigg|_0 \qquad (16.64c)$$
$$= 2U''(a)$$

The coordinates q_s and q_{s+1} are contained in only one term so that

$$a_{s,s+1} = \frac{\partial^2 U}{\partial q_s \partial q_{s+1}}(a + q_{s+1} - q_s)\bigg|_0 = -U''(a) \qquad (16.65c)$$
$$a_{s-1,s} = -U''(a)$$

where the second relation follows from similar considerations. If we define

$$U''(a) \equiv \alpha \qquad (16.66c)$$

then

$$a_{ss} = 2\alpha$$
$$a_{s,s+1} = a_{s-1,s} = -\alpha$$

(16.67c)

All other values of $a_{ss'}$ are zero, since only nearest neighbor interactions have been retained (recall $a_{ss'} = a_{s's}$ is symmetric). The potential is obtained by combining Eqs.(16.60c - 16.67c)

$$U = \sum_{s=1}^{N} \left[-\frac{\alpha}{2} q_s q_{s-1} + \alpha q_s^2 - \frac{\alpha}{2} q_s q_{s+1} \right]$$

(16.68c)

Now introduce *normal coordinates* as in Eq.(16.17c)

$$q_s = \sum_{l=1}^{N} c_{sl} q_l' \qquad ; \ s = 1, 2, \ldots, N$$

(16.69c)

The coefficients in this expansion are obtained from the eigenvectors of the normal mode problem in Eq.(16.27c)

$$\sum_{s'=1}^{N} a_{ss'} c_{s'}^{(l)} = \lambda_l c_s^{(l)} \qquad ; \ s = 1, 2, \ldots, N$$

(16.70c)

according to Eq.(16.33c)

$$c_{sl} = c_s^{(l)} \qquad ; \ s, l = 1, 2, \ldots, N$$

(16.71c)

Retention of the only non-vanishing contributions in Eq.(16.67c) reduces the eigenvalue equations for the linear chain to the following explicit form

$$-\alpha c_{s-1}^{(l)} + 2\alpha c_s^{(l)} - \alpha c_{s+1}^{(l)} = \lambda_l c_s^{(l)} \qquad ; \ s = 1, 2, \ldots, N$$

(16.72c)

where to ensure the fixed endpoints of Eqs.(16.59c) the following condition is imposed on the transformation coefficients in Eqs.(16.69c)

$$c_0^{(l)} = c_{N+1}^{(l)} \equiv 0 \qquad ; \ \text{fixed endpoints}$$

(16.73c)

Equations (16.72c) are a set of linear homogeneous equations for the $\{c_s; s = 1, 2, \ldots, N\}$ and the vanishing of the determinant of coefficients yields the eigenvalues. The required determinant can be evaluated analytically in this case†; however, Eqs.(16.72c) are also finite-difference

---

† See A. L. Fetter and J. D. Walecka, *Theoretical Mechanics of Particles and Continua*, McGraw-Hill, New York (1980).

equations of a relatively simple form, and it is convenient for subsequent applications to seek solutions with a somewhat different procedure. Let us look for solutions in the form of "plane waves"

$$c_s^{(l)} \sim e^{i\kappa_l s} \tag{16.74c}$$

Because the matrix of coefficients $a_{ss'} = a_{s's}$ is real (and symmetric) and the eigenvalues are real (Eq.16.36c), the solution to Eqs.(16.72c) which we seek can then be obtained from either the real or imaginary part of this expression

$$c_s^{(l)} \sim \begin{cases} Re(e^{i\kappa_l s}) & = & \cos(\kappa_l s) \\ Im(e^{i\kappa_l s}) & = & \sin(\kappa_l s) \end{cases} \tag{16.75c}$$

Substitution of Eq.(16.74c) into Eqs.(16.72c) leads to the following expression:

$$-\alpha e^{i\kappa_l(s-1)} + 2\alpha e^{i\kappa_l s} - \alpha e^{i\kappa_l(s+1)} = \lambda_l e^{i\kappa_l s} \quad ; \; s = 1, 2, \ldots, N \tag{16.76c}$$

which can be simplified to

$$\begin{aligned} \lambda_l &= -\alpha e^{-i\kappa_l} + 2\alpha - \alpha e^{i\kappa_l} \\ &= 2\alpha(1 - \cos \kappa_l) \end{aligned} \tag{16.77c}$$

These are the *eigenvalues* and define the normal mode frequencies through Eq.(16.51c)

$$\lambda_l = m\omega_l^2 \tag{16.78c}$$

The normal coordinates perform simple harmonic motion at these frequencies according to Eqs.(16.52c)

$$q_l'(t) = A_l \cos(\omega_l t + \phi_l) \quad ; \; l = 1, 2, \ldots, N \tag{16.79c}$$

A combination of Eqs.(16.77c) and (16.78c) relates the quantity κ_l appearing in the plane wave solution and the normal mode frequencies

$$\omega_l^2 = \frac{4\alpha}{m} \sin^2\left(\frac{\kappa_l}{2}\right) \quad ; \; l = 1, 2, \ldots, N \tag{16.80c}$$

As yet, however, we have no requirements on κ_l ; these will be provided by the *boundary conditions* of the problem.

We must now impose the boundary conditions of Eqs.(16.59c) which are incorporated through the conditions of Eqs.(16.73c)

$$\begin{aligned} c_0^{(l)} &= 0 \\ c_{N+1}^{(l)} &= 0 \quad ; \; \text{fixed endpoints} \end{aligned} \tag{16.81c}$$

The first of these conditions is met by retaining only the $\sin(\kappa_l s)$ terms in Eqs.(16.75c). The second is met by retaining only those values of κ_l for which the term $\sin(\kappa_l s)$ vanishes at the other endpoint

$$\sin \kappa_l(N+1) = 0 \qquad (16.82c)$$

This condition implies

$$\kappa_l(N+1) = l\pi \qquad ; \ l = 1, 2, \ldots, N \qquad (16.83c)$$

or

$$\kappa_l = \frac{l\pi}{N+1} \qquad ; \ l = 1, 2, \ldots, N \qquad (16.84c)$$

The general analysis of the previous section implies that there must be precisely N eigenvalues for this problem, and we have found N distinct eigenvalues. All other choices of l in Eq.(16.84c) must therefore repeat one of the solutions which has already been obtained (Problem 16.c4). The coefficients which define the transformation to normal coordinates for the linear chain with fixed endpoints are thus given by

$$c_s^{(l)} = c_l \sin \kappa_l s = c_l \sin \left(\frac{l\pi s}{N+1}\right) \qquad ; \ s = 1, 2, \ldots, N \qquad (16.85c)$$

The overall constant must be determined by matching the orthonormality condition of Eq.(16.32c)

$$\sum_{s=1}^{N} c_s^{(l)} c_s^{(p)} = \delta_{lp} \qquad (16.86c)$$

The required finite sums are evaluated explicitly in Appendix D. The result is that

$$\sum_{s=1}^{N} \sin \left(\frac{l\pi s}{N+1}\right) \sin \left(\frac{p\pi s}{N+1}\right) = 0 \qquad \text{for } l \neq p$$

$$= \frac{N+1}{2} \quad \text{for } l = p \qquad (16.87c)$$

The solutions are indeed orthogonal, and the required normalization condition is

$$c_l^2 \left(\frac{N+1}{2}\right) = 1 \qquad (16.88c)$$

which implies that, with an appropriate choice of overall phase, the normalization constants are given by

$$c_l = \left(\frac{2}{N+1} \right)^{\frac{1}{2}} \tag{16.89c}$$

The required transformation to normal coordinates in Eq.(16.69c) has thus been obtained in the form

$$q_s = \left(\frac{2}{N+1} \right)^{\frac{1}{2}} \sum_{l=1}^{N} q_l' \sin \left(\frac{l\pi s}{N+1} \right) \quad ; s = 1, 2, \ldots, N \tag{16.90c}$$

Let us now change the notation slightly. Recall that the equilibrium position of the sth atom is given by

$$(x_s)_0 = sa \equiv x \tag{16.91c}$$

Define the wavenumber according to

$$\kappa_l s \equiv k_l x = k_l sa \tag{16.92c}$$

and label the coordinate q not with a discrete index s denoting the atom, but by the position x defined through Eq.(16.91c). Thus

$$q_s \longrightarrow q(x) \tag{16.93c}$$

The transformation to normal coordinates can then be written as

$$q(x) = \left(\frac{2}{N+1} \right)^{\frac{1}{2}} \sum_{k} q'(k) \sin kx \tag{16.94c}$$

where the sum on k runs over the discrete values

$$k_l = \frac{\kappa_l}{a} = \frac{l\pi}{(N+1)a} = \frac{l\pi}{L} \quad ; l = 1, 2, \ldots, N \tag{16.95c}$$

Here

$$L \equiv (N+1)a \tag{16.96c}$$

is the length of the chain (Figure 16.3c). Each normal coordinate in Eq.(16.94c) is now labelled with the wave number of Eq.(16.95c) according to

$$q_l' \longrightarrow q'(k_l) \tag{16.97c}$$

The normal mode frequencies are given by Eqs.(16.80c) and (16.95c)

$$\omega_l = 2\left(\frac{\alpha}{m}\right)^{\frac{1}{2}}\left|\sin\left(\frac{\kappa_l}{2}\right)\right| = 2\left(\frac{\alpha}{m}\right)^{\frac{1}{2}}\left|\sin\left(\frac{k_l a}{2}\right)\right| \qquad (16.98c)$$

where the allowed values of the wavenumbers are indicated in Eqs.(16.95c). With the labelling of Eq.(16.97c), this relation can be reexpressed as

$$\omega(k) = 2\left(\frac{\alpha}{m}\right)^{\frac{1}{2}}\sin\left(\frac{ka}{2}\right) \qquad (16.99c)$$

This is the *dispersion relation for the linear chain with fixed endpoints*. It is sketched in Figure 16.5c.

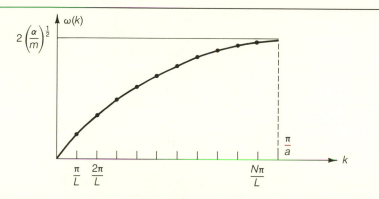

Figure 16.5c Dispersion relation for the linear chain with fixed endpoints given by Eqs.(16.99c) and (16.95c).

To summarize the development up to here, the dispersion relation for the linear chain is given by

$$\omega(k) = \omega^* \sin\left(\frac{ka}{2}\right)$$

$$\omega^* \equiv 2\left(\frac{\alpha}{m}\right)^{\frac{1}{2}} \qquad (16.100c)$$

The allowed values of the wavenumber appearing in this expression are determined from the boundary condition of fixed endpoints

$$k = \frac{\pi}{L}, \frac{2\pi}{L}, \ldots, k^* \qquad ; \text{ fixed endpoints} \qquad (16.101c)$$

The maximum wavenumber is given by

$$k^* = \frac{N\pi}{L} = \frac{N\pi}{(N+1)a} = \frac{\pi}{a}\left(1 - \frac{1}{N+1}\right) = \pi\left(\frac{1}{a} - \frac{1}{L}\right) \qquad (16.102c)$$

Recall that the general relation between wavelength and wavenumber is given by the expression

$$k = \frac{2\pi}{\lambda} \qquad (16.103c)$$

The maximum wavenumber thus corresponds to the shortest wavelength. For a chain whose length is large compared to the interparticle spacing, the maximum wavenumber and corresponding normal mode frequency take the approximate form

$$k_{\mathrm{max}} = k^* \cong \frac{\pi}{a}$$

$$\omega_{\mathrm{max}} \cong \omega^* = 2\left(\frac{\alpha}{m}\right)^{\frac{1}{2}} \qquad (16.104c)$$

The position x here takes the discrete values $x = sa$ as indicated in Eq.(16.91c) and Figure 16.3c, and the longitudinal displacement of each particle is given by

$$q(x) = \left(\frac{2}{N+1}\right)^{\frac{1}{2}} \sum_k q'(k) \, \sin \, kx \qquad (16.105c)$$

where the sum goes over the set of discrete values in Eq.(16.101c). The normal coordinates $q'(k)$ are the amplitudes of the normal mode displacements

$$c_s^{(l)} \longrightarrow c_x(k) = \left(\frac{2}{N+1}\right)^{\frac{1}{2}} \sin \, kx \qquad (16.106c)$$

Since the generalized coordinates of the problem have been transformed from $q(x) \longrightarrow q'(k)$, the generalized momenta must be similarly transformed. The required relation is given by Eqs.(16.41c) and (16.42c)

$$p_s = \sum_{l=1}^{n} \frac{\partial q_s}{\partial q'_l} p'_l \qquad (16.107c)$$

In the notation of Eqs.(16.93c) and (16.97c) this relation becomes

$$p(x) = \sum_k \frac{\partial q(x)}{\partial q'(k)} p'(k) = \left(\frac{2}{N+1}\right)^{\frac{1}{2}} \sum_k p'(k) \, \sin \, kx \qquad (16.108c)$$

The transformations to normal coordinates and momenta in Eqs.(16.105c) and (16.108c) now have precisely the same form. In this same notation, the hamiltonian of Eq.(16.48c) is given by

$$H = \sum_k \left[\frac{p'(k)^2}{2m} + \frac{1}{2}m\omega^2(k)q'(k)^2 \right] \qquad (16.109c)$$

where the sum again goes over the set of discrete values in Eq.(16.101c). This H represents a system of N uncoupled simple harmonic oscillators, each of which oscillates at the normal mode frequency $\omega(k)$; in this notation the solution to Hamilton's equations in Eq.(16.52c) is given by

$$q'(k) = A(k) \ \cos \ [\omega(k)t + \phi(k)] \qquad (16.110c)$$

The normal mode displacements in Eq.(16.106c) represent standing waves, which oscillate independently with the amplitudes of Eq.(16.110c). The minimum wave number is obtained for $l = 1$, where the displacements are given by

$$c_x \left(\frac{\pi}{L} \right) = \left(\frac{2}{N+1} \right)^{\frac{1}{2}} \sin \left(\frac{\pi x}{L} \right) \qquad (16.111c)$$

The maximum wavenumber is obtained for $k = k^*$ where the displacements are given by Eqs.(16.102c) and (16.91c) as

$$c_x(k^*) = \left(\frac{2}{N+1} \right)^{\frac{1}{2}} \sin \left[\pi x \left(\frac{1}{a} - \frac{1}{L} \right) \right]$$

$$= \left(\frac{2}{N+1} \right)^{\frac{1}{2}} \sin \left(\pi s - \frac{\pi x}{L} \right) = \left(\frac{2}{N+1} \right)^{\frac{1}{2}} (-1)^{s+1} \ \sin \ \frac{\pi x}{L}$$
$$(16.112c)$$

These two cases are sketched in Figure 16.6c. In the first case, the atoms all oscillate in the same direction with a wavelength of displacements

$$\lambda_{\max} = \frac{2\pi}{k_{\min}} = 2L \qquad (16.113c)$$

In the second mode, with the maximum wavenumber and shortest wavelength

$$\lambda_{\min} = \frac{2\pi}{k_{\max}} \cong 2a \qquad (16.114c)$$

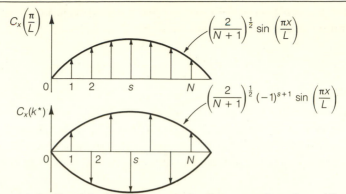

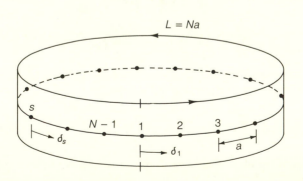

Figure 16.6c Particle displacements in the normal modes with smallest and largest wavenumbers for the linear chain. Here $x = sa$.

the particles move away from their neighbors. Since this mode has the highest frequency, it has the highest energy $\hbar\omega$ in the quantum theory.

Periodic Boundary Conditions. If instead of being held fixed, the two ends of the linear chain are joined as indicated in Figure 16.7c, one has a system described with *periodic boundary conditions.*

In this case, both solutions must be retained in Eqs.(16.75c). The solutions in Eq.(16.75c) are *linearly independent*, and it is evident from

Figure 16.7c Linear chain with ends joined. The system is free to perform longitudinal oscillations; it is described with periodic boundary conditions.

Eq.(16.72c) that any linear combination of the solutions found in Eqs.(16.75c) will again provide a solution; they provide a complete *basis* of solutions to the second-order finite difference Eq.(16.72c). Thus we can write the general form of the transformation to normal coordinates for each l as

$$c_s^{(l)} q_l' \equiv c_1^{(l)} q_{1l}' \cos \kappa_l s + c_2^{(l)} q_{2l}' \sin \kappa_l s \quad ; \quad s = 1, 2, \ldots, N \quad (16.115c)$$

The general transformation to normal coordinates is then obtained by summing this expression over all l as in Eq.(16.69c).

$$q_s = \sum_{l=1}^{N} \left[c_1^{(l)} q_{1l}' \cos \kappa_l s + c_2^{(l)} q_{2l}' \sin \kappa_l s \right] \quad ; \quad s = 1, 2, \ldots, N \quad (16.116c)$$

With the definitions

$$
\begin{aligned}
c_1^{(l)} q_{1l}' &\longrightarrow c_1(k) q_1'(k) \\
c_2^{(l)} q_{2l}' &\longrightarrow c_2(k) q_2'(k)
\end{aligned}
\qquad (16.117c)
$$

the general transformation to normal coordinates in the case of periodic boundary conditions takes the form

$$q(x) = \sum_k [c_1(k) q_1'(k) \cos kx + c_2(k) q_2'(k) \sin kx] \qquad (16.118c)$$

where $c_1(k)$ and $c_2(k)$ are normalization constants to be determined from the orthonormality relations. We now impose the boundary conditions that the physical situation repeat itself after a length $L = Na$. The transformation coefficients must then satisfy

$$\left\{ \begin{matrix} \sin \\ \cos \end{matrix} \right\} [k(x + L)] = \left\{ \begin{matrix} \sin \\ \cos \end{matrix} \right\} [kx] \qquad (16.119c)$$

This implies that in the case of *periodic boundary conditions* the wavenumbers must be given by †

$$kL = 2\pi n \quad ; \quad n = 1, 2, \ldots, \frac{N-1}{2} \qquad (16.120c)$$

There is in addition one non-vanishing mode with $k = 0$. This value was not allowed in the case of a linear chain with fixed endpoints; it corresponds to a uniform translation in the case of periodic boundary conditions ($\cos kx = 1$). We can immediately check that the total number

---

† It is assumed here that N is odd, so that $(N - 1)/2$ is an integer.

of modes is correct. There are two linearly independent modes with $k \neq 0$ in Eq.(16.118c) and one additional mode with $k = 0$. Thus

$$\# \text{ of modes} = 2\left(\frac{N-1}{2}\right) + 1 = N \qquad (16.121c)$$

All other solutions to the periodicity condition of Eq.(16.119c) must reproduce one of the modes we have retained above (Problem 16.c5). The normal mode frequencies are again obtained from Eq.(16.100c)

$$\omega(k) = \omega^* \left| \sin\left(\frac{ka}{2}\right) \right| \qquad (16.122c)$$

The maximum wavenumber is given by

$$\omega^* = 2\left(\frac{\alpha}{m}\right)^{\frac{1}{2}} \qquad (16.123c)$$

The orthonormality condition of Eq.(16.32c) must again be satisfied. The required finite sums are evaluated in Appendix D.

$$\left(\frac{2}{N}\right) \sum_{s=1}^{N} \sin\left(\frac{2\pi ns}{N}\right) \sin\left(\frac{2\pi ms}{N}\right) = \delta_{nm}$$

$$\left(\frac{2}{N}\right) \sum_{s=1}^{N} \cos\left(\frac{2\pi ns}{N}\right) \cos\left(\frac{2\pi ms}{N}\right) = \delta_{nm} \qquad (16.124c)$$

$$\left(\frac{2}{N}\right) \sum_{s=1}^{N} \sin\left(\frac{2\pi ns}{N}\right) \cos\left(\frac{2\pi ms}{N}\right) = 0$$

The normalization constants are therefore

$$c_1(k) = c_2(k) = \left(\frac{2}{N}\right)^{\frac{1}{2}} \qquad (16.125c)$$

The required transformation to normal coordinates and momenta is thus given by

$$q(x) = \left(\frac{2}{N}\right)^{\frac{1}{2}} \sum_{k} [q_1'(k) \ \cos \ kx + q_2'(k) \ \sin \ kx] \qquad (16.126c)$$

$$p(x) = \left(\frac{2}{N}\right)^{\frac{1}{2}} \sum_{k} [p_1'(k) \ \cos \ kx + p_2'(k) \ \sin \ kx] \qquad (16.127c)$$

where the sum goes over the finite set of values in Eq.(16.120c). The hamiltonian of Eq.(16.48c) now takes the form

$$H = \frac{P^2}{2M} + \sum_k \left\{ \frac{1}{2m}[p_1'(k)^2 + p_2'(k)^2] + \frac{m\omega^2(k)}{2}[q_1'(k)^2 + q_2'(k)^2] \right\}$$

(16.128c)

Here $M = Nm$ and the contribution to the kinetic energy from the center-of-mass motion of the chain, the $k = 0$ term, has been explicitly separated.† The particle displacements and momenta in Eqs.(16.126c-16.127c) correspondingly no longer contain this uniform translation.

The normal coordinates $q_1'(k), q_2'(k)$ and momenta $p_1'(k), p_2'(k)$ in Eqs.(16.126c) and (16.127c) are independent quantities, and Hamilton's equations imply uncoupled simple harmonic oscillators.

$$\dot{q}_1'(k) = \frac{\partial H}{\partial p_1'(k)} = \frac{p_1'(k)}{m}$$
$$-\dot{p}_1'(k) = \frac{\partial H}{\partial q_1'(k)} = m\omega^2(k)q_1'(k)$$

(16.129c)

$$\dot{q}_2'(k) = \frac{\partial H}{\partial p_2'(k)} = \frac{p_2'(k)}{m}$$
$$-\dot{p}_2'(k) = \frac{\partial H}{\partial q_2'(k)} = m\omega^2(k)q_2'(k)$$

(16.130c)

Thus each normal coordinate can be assigned a distinct amplitude and phase

$$q_1'(k) = A_1(k) \, \cos\, [\omega(k)t + \phi_1(k)]$$
$$q_2'(k) = A_2(k) \, \cos\, [\omega(k)t + \phi_2(k)]$$

(16.131c)

$$p_1'(k) = -m\omega(k)A_1(k) \, \sin\, [\omega(k)t + \phi_1(k)]$$
$$p_2'(k) = -m\omega(k)A_2(k) \, \sin\, [\omega(k)t + \phi_2(k)]$$

(16.132c)

The contribution to the particle displacements from each of the terms in Eq.(16.126c) is evidently determined by *standing waves*. Suppose instead one seeks to describe the motions in terms of *running waves*. How can this be accomplished? Introduce the quantities q(k) and p(k) in the following manner

$$q(x) \equiv \frac{1}{\sqrt{N}} \sum_{k=-k^*}^{k^*} {}' \left[q(k) \, \cos\, kx - \frac{p(k)}{m\omega(k)} \, \sin\, kx \right]$$

(16.133c)

_____

† Define $p_1'(0) \equiv P/\sqrt{N}$ and note that $\omega^2(0)q_1'(0) \equiv 0$.

$$p(x) \equiv \frac{1}{\sqrt{N}} \sum_{k=-k^*}^{k^*} {}' [p(k) \ \cos \ kx + m\omega(k)q(k) \ \sin \ kx] \quad (16.134c)$$

Here $k^* \equiv \pi(N-1)/L$ and the prime on the summation indicates that the $k = 0$ term is explicitly omitted. We here and henceforth use the convention that $\omega(k)$ is the positive quantity defined by

$$\omega(k) \equiv \omega^* \left| \sin\left(\frac{ka}{2}\right) \right| \quad (16.135c)$$

A change of dummy summation variables in Eq.(16.133c) yields

$$q(x) = \frac{1}{\sqrt{N}} \sum_{k=2\pi/L}^{k^*} \Big([q(k) \ \cos \ kx + q(-k) \ \cos \ (-kx)]$$
$$- \frac{1}{m\omega(k)}[p(k) \ \sin \ kx + p(-k) \ \sin(-kx)] \Big) \quad (16.136c)$$

and Eq.(16.126c) permits the identification

$$\sqrt{2}q_1'(k) = q(k) + q(-k)$$
$$\sqrt{2}q_2'(k) = -\frac{1}{m\omega(k)}[p(k) - p(-k)] \quad (16.137c)$$

A similar calculation with Eq.(16.134c) yields

$$\sqrt{2}p_1'(k) = p(k) + p(-k)$$
$$\sqrt{2}p_2'(k) = m\omega(k)[q(k) - q(-k)] \quad (16.138c)$$

These relations can be inverted to give

$$q(k) = \frac{1}{\sqrt{2}}\left[q_1'(k) + \frac{p_2'(k)}{m\omega(k)}\right]$$
$$q(-k) = \frac{1}{\sqrt{2}}\left[q_1'(k) - \frac{p_2'(k)}{m\omega(k)}\right] \quad (16.139c)$$
$$p(k) = \frac{1}{\sqrt{2}}[p_1'(k) - m\omega(k)q_2'(k)]$$
$$p(-k) = \frac{1}{\sqrt{2}}[p_1'(k) + m\omega(k)q_2'(k)] \quad (16.140c)$$

It is readily established from Eqs.(16.129c) and (16.130c) that these new quantities satisfy Hamilton's equations

$$\dot{q}(k) = p(k)/m$$
$$-\dot{p}(k) = m\omega^2(k)q(k)$$

(16.141c)

$$\dot{q}(-k) = p(-k)/m$$
$$-\dot{p}(-k) = m\omega^2(k)q(-k)$$

(16.142c)

so the transformation to the new quantities is a *canonical transformation*. We started with $(N-1)$ values of $q_1'(k)$, $q_2'(k)$, $p_1'(k)$, $p_2'(k)$ and we now have $2(N-1)$ values of $q(k)$ and $p(k)$. The hamiltonian in Eq.(16.128c) can be written in terms of these new quantities through the use of the equality

$$\frac{1}{2m}[p_1'(k)^2 + p_2'(k)^2] + \frac{1}{2}m\omega^2(k)[q_1'(k)^2 + q_2'(k)^2]$$
$$= \frac{1}{2m}p^2(k) + \frac{1}{2}m\omega^2(k)q^2(k) + (k \rightleftharpoons -k)$$

(16.143c)

A change in dummy summation variables in the contribution of the last term then yields

$$H = \frac{P^2}{2M} + {\sum_k}' \left[\frac{p^2(k)}{2m} + \frac{1}{2}m\omega^2(k)q^2(k) \right]$$

(16.144c)

where the sum now goes over the discrete set

$$k = \frac{2\pi n}{L} \quad ; \quad n = \pm 1, \pm 2, \ldots, \pm\frac{1}{2}(N-1)$$

(16.145c)

and the prime denotes that the $k = 0$ term is again to be omitted in the sum. The hamiltonian of Eq.(16.144c) again represents a sum of uncoupled harmonic oscillators, and Hamilton's equations for the $q(k)$ and $p(k)$ yield the solutions

$$q(k) = A(k) \cos [\omega(k)t + \phi(k)]$$ (16.146c)
$$p(k) = -m\omega(k)A(k) \sin [\omega(k)t + \phi(k)]$$ (16.147c)

Substitution of these relations in the defining Eq.(16.133c) gives

$$q(x) = \frac{1}{\sqrt{N}} {\sum_k}' \left\{ A(k) \cos [\omega(k)t + \phi(k)] \cos kx \right.$$
$$\left. - \frac{1}{m\omega(k)} (-A(k)m\omega(k) \sin [\omega(k)t + \phi(k)]) \sin kx \right\}$$

(16.148c)

which may be rewritten as

$$q(x) = \frac{1}{\sqrt{N}} \sum_k {}'A(k) \cos{[kx - \omega(k)t - \phi(k)]} \qquad (16.149c)$$

Similarly

$$p(x) = \frac{1}{\sqrt{N}} \sum_k {}'[m\omega(k)A(k)] \sin{[kx - \omega(k)t - \phi(k)]} \qquad (16.150c)$$

Equations (16.149c) and (16.150c) are superpositions of *running waves*. Let x represent the location of the point of constant phase. Then

$$kx - \omega(k)t - \phi(k) \equiv k[x - v_p(k)t] - \phi(k) = \text{const.} \qquad (16.151c)$$

Differentiation of this relation with respect to time shows that $v_p(k)$ is the *phase velocity* of the wave. Note that

$$v_p(k) \equiv \frac{\omega(k)}{k} \begin{matrix} > \\ < \end{matrix} 0 \qquad (16.152c)$$

so that the running waves travel in both the positive and negative x direction. Furthermore

$$|v_p(k)| = \frac{\omega(k)}{|k|} = \frac{\omega^* \left| \sin\left(\frac{ka}{2}\right) \right|}{|k|} \qquad (16.153c)$$

This phase velocity is sketched in Figure 16.8c. For long wavelengths $|ka| << 1$ one has

$$|v_p(k)| \cong \frac{\omega^* |ka/2|}{|k|} = \frac{\omega^* a}{2} \qquad (16.154c)$$

The phase velocity is thus independent of k in this limit. For long wavelengths the system of particles looks like an *elastic continuum* and

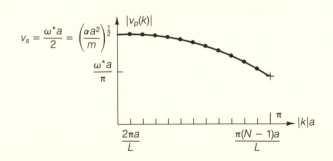

Figure 16.8c Phase velocity versus k for running waves in the linear chain with periodic boundary conditions.

one has *sound waves* travelling in the medium. The longitudinal chain
has a well-defined sound velocity in this limit

$$v_s = 2\left(\frac{\alpha}{m}\right)^{\frac{1}{2}}\frac{a}{2} = \left(\frac{\alpha a^2}{m}\right)^{\frac{1}{2}} \qquad (16.155c)$$

Here Eq.(16.100c) has been employed.

Three-Dimensional Solid. Consider now a real three-dimensional solid.
Assume identical atoms and a simple cubic structure for the lattice with
a lattice spacing of a in each direction. The situation is illustrated in
Figure 16.9c.

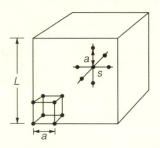

Figure 16.9c A three-dimensional solid with a simple cubic structure.

The large cubic crystal will be assumed to have a length L in each
direction. Instead of fixed boundaries and standing waves in the crystal,
periodic boundary conditions will be assumed. One imagines that a very
large sample is divided into identical large cubes and that the physical
conditions repeat in each large cube. The total number of atoms N in
the cube is given by

$$N = \left(\frac{L}{a}\right)^3 \qquad (16.156c)$$

This follows since there is one atom per unit cell in the simple cubic
structure.

Consider an atom at the equilibrium position **r**. This atom can be
displaced from its equilibrium position by an amount **q(r)** as indicated in
Figure 16.10c. The displacement **q(r)** forms a *vector field*.

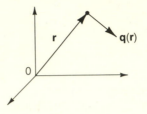

Figure 16.10c The displacement $\mathbf{q}(\mathbf{r})$ of an atom at the equilibrium position $\mathbf{r}$ in the three-dimensional crystal.

The goal is to extend the normal-coordinate expansion of the displacements and momenta in Eqs.(16.146c) and (16.147c) to three dimensions. In analogy with our discussion of the longitudinal oscillations of the linear chain, a first attempt at a three-dimensional extension of Eqs.(16.133c) and (16.134c) might take the following form

$$\mathbf{q}(\mathbf{r}) \cong \frac{1}{\sqrt{N}} \sum_{\mathbf{k}} {}' \left[q(\mathbf{k}) \ \cos \ (\mathbf{k} \cdot \mathbf{r}) - \frac{p(\mathbf{k})}{m\omega(\mathbf{k})} \ \sin \ (\mathbf{k} \cdot \mathbf{r}) \right] \qquad (16.157c)$$

$$\mathbf{p}(\mathbf{r}) \cong \frac{1}{\sqrt{N}} \sum_{\mathbf{k}} {}' \left[p(\mathbf{k}) \ \cos \ (\mathbf{k} \cdot \mathbf{r}) + m\omega(\mathbf{k}) q(\mathbf{k}) \ \sin \ (\mathbf{k} \cdot \mathbf{r}) \right] \qquad (16.158c)$$

The term with $\mathbf{k} = 0$, corresponding to a translation of the crystal, is again to be omitted from the sums. We demand periodicity in all three directions

$$x_i \longrightarrow x_i + L \qquad ; i = x, y, z \qquad (16.159c)$$

This implies that the wavenumbers must satisfy the *periodic boundary conditions*

$$k_i = \frac{2\pi n_i}{L} \qquad ; i = x, y, z \qquad (16.160c)$$
$$n_i = \text{integer}$$

Each term in the sum represents a running wave, and as with the linear chain, all wavenumbers shall be included which satisfy the conditions

$$-k^* \leq k_i \leq k^* \qquad ; i = x, y, z \qquad (16.161c)$$

where the maximum wavenumber is taken from the expression for the linear chain in Eq.(16.102c)

$$k^* = \pi \left(\frac{1}{a} - \frac{1}{L} \right) \qquad (16.162c)$$

In the three-dimensional wavenumber space, or k-space, each allowed wavenumber is represented by a point, and one again finds a simple cubic lattice of allowed wavenumbers (Figure 16.11c). This is known as the *reciprocal lattice*.

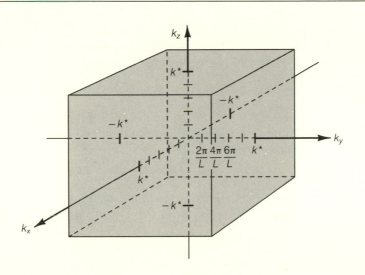

Figure 16.11c Reciprocal lattice of allowed wavenumbers for a simple cubic crystal with periodic boundary conditions. The modes in the first Brillouin zone are indicated.

The region in k-space defined by the conditions in Eqs.(16.161c) is known as the *first Brillouin zone*. The elementary cell in the reciprocal lattice has length $2\pi/L$. Equations (16.160c) imply that there is one distinct mode per cell, and the total number of cells in the first Brillouin zone is

$$\# \text{ of cells} = \frac{(2k^*)^3}{(2\pi/L)^3} \cong \frac{(2\pi/a)^3}{(2\pi/L)^3} = \left(\frac{L}{a} \right)^3 \qquad (16.163c)$$

But this only gives us a total of N normal modes and a corresponding number of N normal coordinates. For a three-dimensional solid there must be a total number of $3N$ degrees of freedom. How does one get around this? It must be realized that for each value of $\mathbf{k}$ there are three independent unit vectors in Figure 16.11c and hence three independent directions. Introduce a complete orthonormal set of unit vectors $\hat{u}(\mathbf{k})$ for each value of the wavenumber $\mathbf{k}$ as indicated in Figure 16.12c.

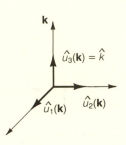

Figure 16.12c Complete set of orthonormal unit vectors for each value of the wavenumber $\mathbf{k}$.

We assign an independent normal coordinate $\mathbf{q}_\gamma(\mathbf{k})$ with $\gamma = 1, 2, 3$ to each of these three directions and write the proper form of Eqs.(16.157c) and (16.158c) as

$$
\mathbf{q}(\mathbf{r}) = \frac{1}{\sqrt{N}} \sum_{\mathbf{k}}' \sum_{\gamma=1}^{3} \hat{u}_\gamma(\mathbf{k}) \Big[q_\gamma(\mathbf{k}) \cos (\mathbf{k} \cdot \mathbf{r})
$$
$$
- \frac{p_\gamma(\mathbf{k})}{m\omega_\gamma(\mathbf{k})} \sin (\mathbf{k} \cdot \mathbf{r}) \Big] \tag{16.164c}
$$

$$
\mathbf{p}(\mathbf{r}) = \frac{1}{\sqrt{N}} \sum_{\mathbf{k}}' \sum_{\gamma=1}^{3} \hat{u}_\gamma(\mathbf{k}) \Big[p_\gamma(\mathbf{k}) \cos (\mathbf{k} \cdot \mathbf{r})
$$
$$
+ m\omega_\gamma(\mathbf{k}) q_\gamma(\mathbf{k}) \sin (\mathbf{k} \cdot \mathbf{r}) \Big] \tag{16.165c}
$$

where the sum over $\mathbf{k}$ goes over the reciprocal lattice defined in Eqs.(16.160 c-16.162c) and shown in Figure 16.11c. The evident three-dimensional generalization of the hamiltonian in Eq.(16.144c) is

$$
H = \frac{P^2}{2M} + \sum_{\mathbf{k}}' \sum_{\gamma} \left[\frac{p_\gamma^2(\mathbf{k})}{2m} + \frac{1}{2} m\omega_\gamma^2(\mathbf{k}) q_\gamma^2(\mathbf{k}) \right] \tag{16.166c}
$$

The magnitude of the phase velocity for each of the running waves is given by Eq.(16.152c)

$$|v_\gamma(\mathbf{k})| = \frac{\omega_\gamma(\mathbf{k})}{|\mathbf{k}|} \qquad (16.167c)$$

To get the dispersion relation $\omega_\gamma(\mathbf{k})$ for the normal modes and the corresponding normal coordinates is, in fact, fairly complicated. If $|\mathbf{k}|a << 1$ and the wavelengths are very long, however, then the system looks like a continuous medium and one can replace $v_\gamma(\mathbf{k}) \longrightarrow v_s$ where v_s is the *velocity of sound* in the medium. The $\hat{u}_\gamma(\mathbf{k})$ in Figure 16.12c consist of a longitudinal vector and two transverse vectors. One can correspondingly speak of the (longitudinal) sound velocity and of transverse sound waves. For many solids, sound propagation is anisotropic with $v_1 \neq v_2 \neq v_3$.

Consider $\mathbf{k}$-space again and assume L very large, that is $L >> a$. In this case the size of the unit cell becomes small, and one can consider a differential volume element $d\mathbf{k}$. The number of modes with distinct $\mathbf{k}$ in this differential volume element is given by

$$\text{\# of modes in volume } d\mathbf{k} = \frac{dk_x dk_y dk_z}{(2\pi/L)^3} \qquad (16.168c)$$

The total number of modes in the first Brillouin zone is then given by

$$\text{\# of modes } =$$

$$\int_{\text{1st Brillouin zone}} d\mathbf{k} \Big/ (2\pi/L)^3 = \frac{L^3}{(2\pi)^3}(2k^*)^3 \cong \left(\frac{L}{a}\right)^3 = N$$

$$(16.169c)$$

which checks the previous answer in Eq.(16.163c).

Through (16.164c-16.166c) we have described a system of $N-1$ simple harmonic oscillators. Each has two coordinates p, q and three polarizations. In addition, there is the motion of the center of mass. The *equipartition theorem* then implies that the total energy is given by

$$E = (N-1)\cdot\frac{3\cdot 2}{2}(k_B T) + \frac{3}{2}(k_B T) = 3N(k_B T) - \frac{3}{2}(k_B T)$$

$$\cong 3Nk_B T \qquad ; \; N >> 1$$

$$(16.170c)$$

The constant volume heat capacity is

$$C_V = \left(\frac{\partial E}{\partial T}\right) = 3Nk_B \qquad (16.171c)$$

and the molar heat capacity is

$$c_v = 3R \qquad (16.172c)$$

Thus one recovers the *Law of Dulong and Petit.*

Although we have worked hard to describe the classical mechanics of the small oscillations of a solid crystal lattice about equilibrium, the basic problem of the disagreement between the observed specific heat and the result predicted by classical statistical mechanics [Figure 16.2c] remains. This paradox can only be resolved within the framework of quantum statistical mechanics where, as we shall see, the normal-mode analysis developed here provides an essential simplification.

This normal-mode analysis also provides a theoretical framework for discussing the (apparently unrelated) problem of black-body radiation, and we proceed to that development.

d. Black-Body Radiation. We now extend the discussion of sound waves in an elastic medium to the case of electromagnetic waves in a cavity. Consider a cavity with reflecting walls and a small hole in one of the walls which permits the radiation spectrum in the cavity to be sampled as illustrated in Figure 16.1d.

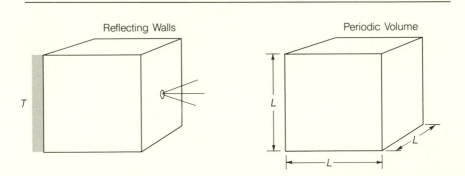

Figure 16.1d Cavity with reflecting walls in equilibrium at temperature T containing black-body radiation and cavity of volume $V = L^3$ with periodic boundary conditions.

In equilibrium at a temperature T, there will be a certain amount of electromagnetic radiation inside of this cavity due to the thermal motion of the particles in the walls. There is absorption and re-emission of the radiation by the particles in the walls, and these processes will bring the entire system to equilibrium. In a *black body* each wave that hits the wall will be absorbed and the system will come to equilibrium as fast as possible. Perfect black bodies do not exist; however, our considerations will involve only the equilibrium properties of the radiation in the cavity and will be independent of the detailed nature of the walls. In fact, we shall consider the radiation in a cavity of volume $V = L^3$ with periodic boundary conditions (Figure 16.1d).

Inside of the cavity one has an electromagnetic field in vacuum. There are two vector fields, the electric field **E** and the magnetic field **H**, and they satisfy *Maxwell's equations*

$$\nabla \cdot \mathbf{E} = 0 \tag{16.1d}$$

$$\nabla \cdot \mathbf{H} = 0 \tag{16.2d}$$

$$\nabla \wedge \mathbf{E} = -\frac{1}{c}\frac{\partial \mathbf{H}}{\partial t} \tag{16.3d}$$

$$\nabla \wedge \mathbf{H} = \frac{1}{c}\frac{\partial \mathbf{E}}{\partial t} \tag{16.4d}$$

Here c is the velocity of light. For plane waves in vacuum, the fields have the spatial configuration illustrated in Figure 16.2d.

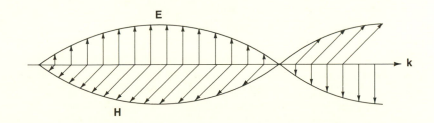

Figure 16.2d Electric and magnetic field configurations for a plane wave with wavenumber **k** in vacuum.

The introduction of a transverse vector potential $\mathbf{A}(\mathbf{r})$ satisfying

$$\nabla \cdot \mathbf{A}(\mathbf{r}) = 0 \tag{16.5d}$$

and related to the electric and magnetic fields by

$$\mathbf{E} = -\frac{1}{c}\frac{\partial \mathbf{A}}{\partial t}$$
$$\mathbf{H} = \nabla \wedge \mathbf{A} \tag{16.6d}$$

permits Maxwell's equations to be recast in the form of the wave equation

$$\nabla^2 \mathbf{A} - \frac{1}{c^2}\frac{\partial^2 \mathbf{A}}{\partial t^2} = 0 \tag{16.7d}$$

This equation possesses plane wave solutions. The appropriate three-dimensional generalization of Eq.(16.149c) then provides an immediate normal mode expansion of the vector field $\mathbf{A}(\mathbf{r})$

$$\mathbf{A}(\mathbf{r}) = c\left(\frac{4\pi}{V}\right)^{\frac{1}{2}} \sum_{\mathbf{k}} \sum_{\gamma=1}^{2} \hat{u}_\gamma(\mathbf{k}) \left\{ A_\gamma(\mathbf{k}) \cos\left[(\mathbf{k}\cdot\mathbf{r}) - \omega_\gamma(\mathbf{k})t - \phi_\gamma(\mathbf{k})\right] \right\}$$

$$\tag{16.8d}$$

Only the *transverse* unit vectors in Figure 16.12c are retained in this expression so that the condition (16.5d) is satisfied. The phase velocity of the waves is the velocity of light

$$c \equiv \frac{\omega_\gamma(\mathbf{k})}{|\mathbf{k}|} \tag{16.9d}$$

If we refer to Figure 16.8c and the discussion of the phase velocity of waves in the linear chain, we observe that in the case of electromagnetic waves in vacuum, there is *no finite length a* associated with the system, and hence one is always effectively in the *long wavelength limit* where there is constant phase velocity and no dispersion. The choice of overall normalization in Eq.(16.8d) will be discussed below.

We again apply *periodic boundary conditions* to the system as in Eqs.(16.159c) and (16.160c).

$$(k_x, k_y, k_z) = \frac{2\pi}{L}(n_x, n_y, n_z) \quad ; \quad n_i = 0, \pm 1, \pm 2, \ldots, \pm\infty$$
$$i = x, y, z \tag{16.10d}$$

All values of the integer set n_x, n_y, n_z are now allowed ; a is zero and hence k^* in Eq.(16.102c) is infinite.

In order to establish the connection to classical statistical mechanics we shall recast Maxwell's equations in *hamiltonian form*. We know from the previous section that the following hamiltonian

$$H = \sum_{\mathbf{k}}{}' \sum_{\gamma=1}^{2} \frac{1}{2} \left[p_\gamma^2(\mathbf{k}) + \omega_\gamma^2(\mathbf{k}) q_\gamma^2(\mathbf{k}) \right] \tag{16.11d}$$

and the normal mode expansion

$$\mathbf{A}(\mathbf{r}) = c \left(\frac{4\pi}{V} \right)^{\frac{1}{2}} \sum_{\mathbf{k}}{}' \sum_{\gamma=1}^{2} \hat{u}_\gamma(\mathbf{k}) \left[q_\gamma(\mathbf{k}) \cos(\mathbf{k} \cdot \mathbf{r}) - \frac{p_\gamma(\mathbf{k})}{\omega_\gamma(\mathbf{k})} \sin(\mathbf{k} \cdot \mathbf{r}) \right]$$

$$\tag{16.12d}$$

will then give rise to Eq.(16.8d) through the use of Hamilton's equations. This is readily established, since

$$\dot{q}_\gamma(\mathbf{k}) = \frac{\partial H}{\partial p_\gamma(\mathbf{k})} = p_\gamma(\mathbf{k})$$

$$-\dot{p}_\gamma(\mathbf{k}) = \frac{\partial H}{\partial q_\gamma(\mathbf{k})} = \omega_\gamma^2(\mathbf{k}) q_\gamma(\mathbf{k}) \tag{16.13d}$$

A combination of these results gives the uncoupled simple harmonic oscillator equations for the normal coordinates

$$\ddot{q}_\gamma(\mathbf{k}) = -\omega_\gamma^2(\mathbf{k}) q_\gamma(\mathbf{k}) \tag{16.14d}$$

The solution to these equations is

$$q_\gamma(\mathbf{k}) = A_\gamma(\mathbf{k}) \cos[\omega_\gamma(\mathbf{k})t + \phi_\gamma(\mathbf{k})]$$

$$p_\gamma(\mathbf{k}) = -\omega_\gamma(\mathbf{k}) A_\gamma(\mathbf{k}) \sin[\omega_\gamma(\mathbf{k})t + \phi_\gamma(\mathbf{k})] \tag{16.15d}$$

The substitution of these relations back into Eq.(16.12d) then immediately yields Eqs.(16.8d). The prime in the sum in Eq.(16.12d) again indicates that the $\mathbf{k} = 0$ term is to be omitted; in the present case the $\mathbf{k} = 0$ term corresponds to static fields. Furthermore, we have the frequency relations

$$\omega_\gamma(\mathbf{k}) = \omega_1(\mathbf{k}) = \omega_2(\mathbf{k}) = \omega(\mathbf{k}) = |\mathbf{k}|c \tag{16.16d}$$

Thus one can consider the system of electromagnetic radiation in a cavity as a collection of uncoupled oscillators with an isotropic frequency distribution.

It is readily established that these arguments are independent of the choice of the overall normalization constant in Eq.(16.12d). The particular choice made here allows the hamiltonian to be identified with the total energy in the electromagnetic field

$$H = \frac{1}{8\pi} \int_V \left(\mathbf{E}^2 + \mathbf{H}^2\right) d\mathbf{r} \qquad (16.17d)$$

when the electric and magnetic fields are calculated from the vector potential in Eq.(16.12d) through the use of Eqs.(16.6d). The demonstration of this result is left as an exercise for the reader (Problem 16.d2).

We now have a hamiltonian for the system of electromagnetic radiation in a cavity, properly expressed in terms of canonical coordinate and momentum variables, and all the previous results of classical statistical mechanics again apply. In particular, the hamiltonian in Eq.(16.11d) is a homogeneous quadratic function of the coordinates and momenta. Each oscillator with wave number $\mathbf{k}$ and two polarizations therefore makes the following contribution to the energy of the system

$$E(\mathbf{k}) = 2 \cdot 2 \cdot \frac{1}{2} k_B T = 2 k_B T \qquad ; \text{ each } \mathbf{k} \qquad (16.18d)$$

To get the total energy of the sytem we must sum over all wavenumbers $\mathbf{k}$. *But there is no cutoff on* $\mathbf{k}$, since there is no underlying lattice as there was in the case of the crystal! Thus the total energy of a black body is infinite in classical statistical mechanics

$$E_{\text{tot}} = \sum_{\mathbf{k}}{}' (2 k_B T) = \infty \quad ! \qquad (16.19d)$$

Once again, the physical reason for this result is that one can have electromagnetic radiation with arbitrarily short wavelength and hence arbitrarily large wavenumber $\mathbf{k}$. One can associate a simple harmonic oscillator with each wavenumber as we have seen, and the equipartition theorem then indicates that each oscillator makes the same contribution to the energy in Eq.(16.18d) *independent of the wavenumber* $\mathbf{k}$. The sum over all $\mathbf{k}$ then diverges. This phenomenon is known as the *ultraviolet catastrophe*. The only way around this dilemma is to again freeze out the high-energy modes where $k_B T \ll \hbar\omega$ by appealing to quantum mechanics.

Consider $\mathbf{k}$-space as depicted in Figure 16.3d. When L is large, sums over the reciprocal lattice can again be replaced by integrals as in Eq.(16.168c). The total number of modes lying within the spherical shell between k and $k + dk$ is given by

$$dn = \text{\# of modes in shell between } k \text{ and } k + dk$$

$$dn = 4\pi k^2 dk \Big/ (2\pi/L)^3 \tag{16.20d}$$

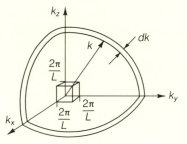

Figure 16.3d Wavenumber space for black-body radiation

The result of the equipartition theorem in Eq.(16.18d) implies that the contribution to the total energy of the modes lying within this shell is given by

$$dE = (2k_BT)dn = (2k_BT) \left(\frac{L}{2\pi} \right)^3 4\pi k^2 dk \tag{16.21d}$$

With the introduction of the frequency through the relation

$$k = \frac{2\pi\nu}{c} \tag{16.22d}$$

this result may be rewritten as

$$dE = \frac{8\pi L^3}{c^3}(k_BT)\nu^2 d\nu \tag{16.23d}$$

If one defines the differential energy density in the cavity by

$$\frac{1}{V}\frac{dE}{d\nu} \equiv \frac{d\epsilon}{d\nu} \tag{16.24d}$$
$$= \text{Energy/unit volume/unit frequency}$$

Then it follows that

$$\frac{d\epsilon}{d\nu} = \frac{8\pi k_B T \nu^2}{c^3} \qquad (16.25d)$$

This is the celebrated *Rayleigh-Jeans law*. This result is sketched in Figure 16.4d, along with the experimentally observed black-body spectrum. The predicted spectrum grows with ν^2; this is the ultraviolet catastrophe. It leads to an infinite total energy density when integrated over all frequencies

$$\int_0^\infty \left(\frac{d\epsilon}{d\nu}\right) d\nu = \infty \qquad ! \qquad (16.26d)$$

In contrast, the experimental density, while in agreement with the Rayleigh-Jeans law at low frequencies (long wavelengths), is seen to fall rapidly to zero for large ν.

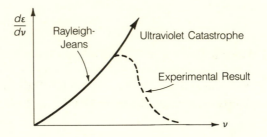

Figure 16.4d Energy per unit volume per unit frequency in a black body: Rayleigh-Jeans law and experimentally observed spectrum.

The observed spectrum is explained by replacing the contribution to the energy of each oscillator by the expression

$$k_B T \longrightarrow \frac{h\nu}{\exp\left(h\nu/k_B T\right) - 1} \qquad (16.27d)$$

This is the *Planck distribution*. We see here that the classical law breaks down. There is no way to explain the experimental curve within the framework of classical statistical mechanics. This led Planck to his revolutionary quantum hypothesis and to the development of quantum mechanics. We turn to this subject in Section VI.

17. Magnetism

In this section we study the phenomena of magnetism.   When a sample is placed in a magnetic field, a magnetization (magnetic dipole moment/volume) is induced in the sample. Suppose the applied field is of the form $\mathbf{H} = \mathcal{H}\hat{x}$, as illustrated in Figure 17.1, and the sample is in contact with a heat bath at a temperature T. Then the induced magnetic dipole moment $\mathbf{M}$ of the sample has the functional form

$$\mathbf{M} = M(\mathcal{H}, T)\hat{x} \tag{17.1}$$

In weak fields, the induced moment of most materials will be proportional to the applied field, and if V is the volume of the sample, then one can define the *magnetic susceptibility* χ by the expression

$$\frac{M}{V} \equiv \chi(T)\mathcal{H} \qquad ; \text{ weak fields} \tag{17.2}$$

If χ is positive so that the induced magnetic moment lies along the applied field, the material is said to be *paramagnetic*. If χ is negative so that the induced moment opposes the field, the material is *diamagnetic*. We start the discussion with a consideration of paramagnetic materials as illustrated in Figure 17.1.

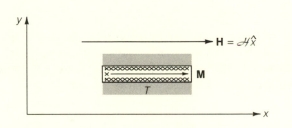

Figure 17.1 Magnetic dipole moment $\mathbf{M}$ induced in a sample at temperature T by an applied external field $\mathbf{H} = \mathcal{H}\hat{x}$.

Consider the magnetic dipole moment $\boldsymbol{\mu}$ induced in a small needle at the position x by a large external permanent dipole magnet placed at the position x' along the x axis with the moment of the external magnet aligned along that axis. The situation is illustrated in Figure 17.2.

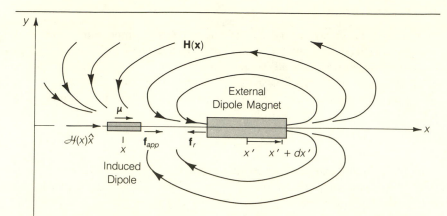

Figure 17.2 Magnetic dipole moment $\boldsymbol{\mu}$ induced in a small needle at the position x by a large external permanent dipole magnet placed at x' with its moment aligned along the x-axis so that it gives rise to the field $\mathbf{H(x)}$.

The field at the sample due to the external permanent dipole magnet will have the form

$$\mathbf{H} = \mathcal{H}(x)\hat{x} \qquad\qquad (17.3a)$$

$$\mathcal{H}(x) = \mathcal{H}(x - x') \sim \frac{1}{(x - x')^3} \qquad ; \text{ dipole field} \qquad (17.3b)$$

Throughout this discussion it is assumed that $x' >> x$. This applied field induces a dipole moment $\boldsymbol{\mu} = \mu\hat{x}$ in the small needle. Since the external applied field in Eq.(17.3) is *inhomogeneous*, there will be a net applied force $\mathbf{f}_{app}$ on the induced dipole which is given by

$$\mathbf{f}_{app} = \mu\frac{d\mathcal{H}(x)}{dx}\hat{x} \equiv -\mu\frac{d\mathcal{H}}{dx'}\hat{x} \qquad\qquad (17.4)$$

The last equality follows from the explicit functional form of the field in Eq.(17.3b). Newton's third law implies that the small needle must exert an equal and opposite force $\mathbf{f}_r$ on the large external permanent dipole (Figure 17.2)

$$\mathbf{f}_r = -\mathbf{f}_{app} = \mu\frac{d\mathcal{H}}{dx'}\hat{x} \qquad\qquad (17.5)$$

Now suppose the external magnet is slowly moved a distance $+dx'$ along the x axis. This results in a change in the applied field $\mathcal{H}(x)$ at the position of the small needle to $\mathcal{H}(x) + d\mathcal{H}(x)$. The work that the system (the induced dipole) does on the external magnet during this change is given by

$$\delta W_{\text{system on external magnet}} = \mathbf{f}_r \cdot d\mathbf{x}' = \mu \left(\frac{d\mathcal{H}}{dx'} \right) dx' \qquad (17.6)$$
$$= \mu d\mathcal{H}(x)$$

The work done *on* the system when the external magnet is moved is defined to be the negative of this quantity

$$\delta W_{\text{external magnet on system}} = -\mu d\mathcal{H}(x) \qquad (17.7)$$

In the situation illustrated in Figure 17.2, the force $\mathbf{f}_r$ must be slightly over-compensated by an external force to move the external magnet a distance dx'. Thus work must be supplied to execute this change. This change decreases the field $H(x)$ at the sample. Thus the work the external magnet does on the system is appropriately positive in Eq.(17.7).

So far only the work done on a single small needle has been considered. If the thermodynamic system under consideration consists of N such subsystems in a volume V with a corresponding density $n = N/V$, then the individual contributions are additive. The total dipole moment of the sample is given by

$$\mathbf{M} = N\boldsymbol{\mu} = M\hat{x} \qquad (17.8)$$

and the dipole moment per unit volume is

$$\mathbf{M}/V = (N/V)\boldsymbol{\mu} = n\boldsymbol{\mu} = n\mu\hat{x} \qquad (17.9)$$

Furthermore, the slow change in the system induced by the movement of the external magnet constitutes a reversible thermodynamic process. The reversible work done *on* the sample when the external field $\mathcal{H}$ is changed slightly by a small movement of the external magnet is therefore

$$\delta W = -M d\mathcal{H} \qquad (17.10)$$

Now it was shown in Eq.(11.3) that the work done on a thermodynamic system in a reversible isothermal process is equal to the change in the free energy of the system

$$\delta W = dF \quad \text{; reversible, isothermal process} \qquad (17.11)$$

Thus we derive the important relation

$$dF = -M d\mathcal{H} \quad \text{; reversible, isothermal process} \qquad (17.12)$$

In *summary*, if classical statistical mechanics and the canonical ensemble are used to calculate the partition function and free energy at a temperature T for a sample of volume V in the presence of an external field $\mathcal{H}$, then the induced magnetization M/V is given by the expression

$$\frac{M}{V} = -\frac{1}{V} \left(\frac{\partial F}{\partial \mathcal{H}} \right)_{V,T} \tag{17.13}$$

To proceed, one must make a model. Assume that the system consists of N point particles interacting through a potential so that the hamiltonian in the absence of a magnetic field is given by

$$H_0 = \sum_{s=1}^{N} \frac{\mathbf{p}_s^2}{2m} + V(\mathbf{r}_1, \mathbf{r}_2, \ldots, \mathbf{r}_N) \tag{17.14}$$

$$\equiv H_0(\mathbf{p}_s, \mathbf{r}_s)$$

Now introduce a static external magnetic field $\mathbf{H}(\mathbf{r})$ through the vector potential $\mathbf{A}(\mathbf{r})$

$$\mathbf{H} = \nabla \wedge \mathbf{A}(\mathbf{r}) \tag{17.15}$$

This will be done by making the following *minimal substitution* in the hamiltonian

$$\mathbf{p}_s \longrightarrow \left[\mathbf{p}_s - \frac{e_s}{c} \mathbf{A}(\mathbf{r}) \right] \tag{17.16}$$

Here e_s is the charge on the sth particle. Thus in a specified static magnetic field

$$H = H_0 \left[\mathbf{p}_s - \frac{e_s}{c} \mathbf{A}(\mathbf{r}_s), \ \mathbf{r}_s \right]$$

$$= \sum_{s=1}^{N} \frac{1}{2m} \left[\mathbf{p}_s - \frac{e_s}{c} \mathbf{A}(\mathbf{r}_s) \right]^2 + V(\mathbf{r}_1, \mathbf{r}_2, \ldots, \mathbf{r}_N) \tag{17.17}$$

The quantity $\mathbf{A}(\mathbf{r})$ appearing in these expressions is fixed; it produces the given external magnetic field $\mathbf{H}(\mathbf{r})$. The dynamical variables are p_i, x_i; $i = 1, 2, \ldots, 3N$. Hamilton's equations then lead to the following equations of motion

$$m\ddot{\mathbf{r}}_s = \mathbf{f}_s \quad ; \ s = 1, 2, \ldots, N \tag{17.18}$$

$$\mathbf{f}_s = -\nabla_s V + \frac{e_s}{c} [\dot{\mathbf{r}}_s \wedge \mathbf{H}] \tag{17.19}$$

(The derivation of these results is left as an exercise for the reader. See Problem 17.6.) The *Lorentz force* is now included in Eq.(17.19), and this serves as the justification for the minimal substitution in Eq.(17.16).

The *partition function* in the presence of the magnetic field is given by

$$Z = \int e^{-\beta H_0[\mathbf{p}_s - e_s \, \mathbf{A}(\mathbf{r}_s)/c, \, \mathbf{r}_s]} \, d\lambda \qquad (17.20)$$

where the phase space volume is defined as usual as

$$d\lambda = \prod_{s=1}^{N} dp_{sx} dp_{sy} dp_{sz} dx_s dy_s dz_s \qquad (17.21)$$

Equation (17.20) looks very complicated; however, it is now possible to eliminate the vector potential $\mathbf{A}(\mathbf{r})$ entirely! Simply make the following change of variables in the integral

$$\mathbf{p}_s \longrightarrow \mathbf{p}'_s \equiv \mathbf{p}_s - \frac{e_s}{c} \mathbf{A}(\mathbf{r}_s) \qquad (17.22)$$

Since $\mathbf{A}(\mathbf{r}_s)$ is a specified function of position and is independent of momentum, one has

$$dp'_{sx} dp'_{sy} dp'_{sz} = dp_{sx} dp_{sy} dp_{sz} \qquad (17.23)$$

and with a simple relabelling of variables, the partition function in Eq.(17.20) takes the form

$$Z = \int e^{-\beta H_0(\mathbf{p}_s, \mathbf{r}_s)} d\lambda \qquad (17.24)$$

So now Z is independent of $\mathcal{H}$ and we have lost the effects of the magnetic field! Equation (11.5) relates the partition function to the free energy

$$F = -\frac{1}{\beta}(\ln Z + \text{ const.}) \qquad (17.25)$$

Hence the free energy F is also independent of the magnetic field $\mathcal{H}$! Thus the induced magnetic dipole moment of the system given by Eq.(17.13) must vanish

$$M = -\left(\frac{\partial F}{\partial \mathcal{H}}\right)_{V,T} = 0 \qquad (17.26)$$

The result that any collection of charged particles in equilibrium in a magnetic field exhibits no magnetic moment in classical statistical mechanics is the *theorem of Van Leeuwen*. One knows that as the magnetic field is turned on, eddy currents are produced in the material giving rise to magnetic moments. In thermodynamic equilibrium, however, the effects of these eddy currents must all die away.

Real physical systems exhibit both paramagnetism and diamagnetism, and the previous general analysis based on classical statistical mechanics left us with no magnetization whatsoever. However, all is not lost. One can achieve a partial description of the behavior of real systems by imposing additional *ad hoc* restrictions, which are only justified later within the framework of quantum mechanics. To illustrate this, we first choose a particular gauge for the vector potential and write

$$\mathbf{A} = \frac{1}{2}(\mathbf{H} \wedge \mathbf{r}) \tag{17.27}$$

If $\mathbf{H}$ is constant, then an elementary calculation establishes that

$$\nabla \wedge \mathbf{A} = \mathbf{H} \tag{17.28}$$

Now place $\mathbf{H}$ along the z-axis so that

$$\mathbf{H} = \mathcal{H}\hat{z}$$
$$A_x = -\frac{1}{2}\mathcal{H}y \quad ; \quad A_y = \frac{1}{2}\mathcal{H}x \quad ; \quad A_z = 0 \tag{17.29}$$

Consider an atom consisting of a heavy nucleus at rest surrounded by Z electrons moving in this magnetic field. The hamiltonian for this atomic subsystem is

$$h = \sum_{s=1}^{Z} \frac{1}{2m}\left[\mathbf{p}_s + \frac{e}{c}\mathbf{A}(\mathbf{r}_s)\right]^2 + V(\mathbf{r}_1, \mathbf{r}_2, \ldots, \mathbf{r}_s) \tag{17.30}$$

where the potential includes the Coulomb interactions between the electrons and with the nucleus. The charge on the electrons is here defined as

$$e_s = -|e| \equiv -e \tag{17.31}$$

The kinetic energy for the sth electron takes the form

$$\frac{1}{2m}\left[\mathbf{p}_s + \frac{e}{c}\mathbf{A}(\mathbf{r}_s)\right]^2 = \frac{1}{2m}\left[\left(p_{sx} - \frac{e}{2c}\mathcal{H}y_s\right)^2 + \left(p_{sy} + \frac{e}{2c}\mathcal{H}x_s\right)^2 + p_{sz}^2\right] \tag{17.32}$$

which can be rewritten as

$$\frac{1}{2m}\left[\mathbf{p}_s + \frac{e}{c}\mathbf{A}(\mathbf{r}_s)\right]^2 = \frac{1}{2m}\mathbf{p}_s^2 + \frac{e\mathcal{H}}{2mc}(x_s p_{sy} - y_s p_{sx}) + \frac{e^2\mathcal{H}^2}{8mc^2}(x_s^2 + y_s^2)$$

$$(17.33)$$

The coefficient in the second term is recognized as the z-component of the angular momentum

$$\ell_{sz} = [\mathbf{r}_s \wedge \mathbf{p}_s]_z = x_s p_{sy} - y_s p_{sx} \qquad (17.34)$$

We define a magnetic moment for the electron by

$$\boldsymbol{\mu}_s \equiv -\frac{e}{2mc}\mathbf{l}_s \qquad (17.35)$$

Then the total magnetic moment of the atom is given by

$$\boldsymbol{\mu} = \sum_{s=1}^{Z}\boldsymbol{\mu}_s = -\frac{e}{2mc}\sum_{s=1}^{Z}\mathbf{l}_s = -\frac{e}{2mc}\mathbf{L} \qquad (17.36)$$

where $\mathbf{L}$ is the total angular momentum of the electrons in the atom. This equation may be written as

$$\boldsymbol{\mu} = \gamma\mathbf{L} \qquad (17.37)$$

where γ is defined as the gyromagnetic ratio of the electron

$$\gamma \equiv -\frac{e}{2mc} \qquad (17.38)$$

A combination of these results permits the hamiltonian of Eq.(17.30) to be rewritten in the form

$$h = h_0 - \boldsymbol{\mu} \cdot \mathbf{H} + \frac{e^2\mathcal{H}^2}{8mc^2}\rho^2 \qquad (17.39)$$

where we have defined

$$\rho^2 \equiv \sum_{s=1}^{Z}(x_s^2 + y_s^2) \qquad (17.40)$$

Now the magnetic moment of the atom averages to zero in classical statistical mechanics, since the ensemble average of the vector angular momentum $\mathbf{L}$ vanishes. It is possible to construct a "classical theory" of magnetism if one makes the *ad hoc* assumption that the magnitude of the angular momentum of an isolated atom is a fixed quantity. Equations

(17.37-17.38) then imply that the magnitude of the magnetic moment of an isolated atom will also be a fixed quantity. Thus one assumes

$$|\mathbf{L}|_{\text{atom}} \equiv \ell = \text{fixed value}$$
$$|\boldsymbol{\mu}|_{\text{atom}} \equiv \mu = |\gamma|\ell = \text{fixed value} \tag{17.41}$$

The partition function will then be found to depend on the magnetic field $\mathcal{H}$, and a realistic expression for the induced magnetization will be obtained. The angular momentum and magnetic moment of an atom are assumed to be "frozen in." The justification comes only from quantum mechanics, where one finds that

$$\ell \sim \hbar \equiv h/2\pi$$
$$\mu \sim e\hbar/2mc \tag{17.42}$$

The *Bohr magneton* is defined by

$$\mu_o \equiv \frac{e\hbar}{2m_e c} = 9.274 \times 10^{-21} \text{erg/Gauss} \tag{17.43}$$

It will also be assumed that the atom has a finite size which is also "frozen in" (although again at this stage we have no apparent reason for this)

$$\rho_{\text{atom}}^2 \equiv \frac{2}{3}Zr^2 = \text{fixed value} \tag{17.44}$$

Quantum mechanics implies that

$$r^2 \sim \left(\frac{\hbar^2}{me^2}\right)^2 \tag{17.45}$$

The *Bohr radius* is defined by

$$a_0 \equiv \frac{\hbar^2}{m_e e^2} = 5.292 \times 10^{-9} \text{cm} \tag{17.46}$$

Now consider N independent atoms. The internal partition function, which contains the entire dependence on the magnetic field, takes the form of Eq.(16.13b)

$$\mathcal{Z}_{\text{int}}(\mathcal{H}) = [z_{\text{int}}(\mathcal{H})]^N \tag{17.47}$$

And, exactly as with the rigid rotor in Eq.(16.25b), the internal partition function will be obtained by integrating the hamiltonian in Eq.(17.39) over all possible orientations of the magnetic moment (Figure 17.3)

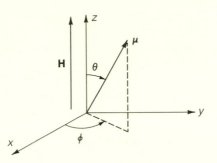

Figure 17.3 Configuration for average over the orientations of the magnetic moment in the partition function for an atom in the external field **H**.

$$z_{\text{int}}(\mathcal{H}) = \frac{1}{4\pi} \int_0^{2\pi} d\phi \int_0^\pi \sin\theta d\theta \; e^{\beta\mu\mathcal{H}\cos\theta} \; e^{-\beta Z e^2 \mathcal{H}^2 r^2 / 12mc^2} \quad (17.48)$$

With no external field there is no dependence of the hamiltonian on the overall orientation of the atom.† Assume that the temperature is high enough so that

$$\beta\mu\mathcal{H} = \frac{\mu\mathcal{H}}{k_B T} \ll 1 \quad (17.49)$$

The first exponential in Eq.(17.48) can then be expanded in a power series and

$$z_{\text{int}}(\mathcal{H}) \cong \frac{1}{4\pi} e^{-\beta Z e^2 r^2 \mathcal{H}^2 / 12mc^2} \int_0^{2\pi} d\phi \int_0^\pi \sin\theta d\theta \left[1 + \beta\mathcal{H}\mu\cos\theta \right.$$
$$\left. + \frac{1}{2}(\beta\mathcal{H}\mu)^2 \cos^2\theta \right]$$
$$(17.50)$$

Since the first factor in Eq.(17.50) is independent of orientation, it has been taken outside the integral. The remaining integrals are immediately performed and

$$z_{\text{int}}(\mathcal{H}) \cong e^{-\beta Z e^2 r^2 \mathcal{H}^2 / 12mc^2} \left[1 + \frac{1}{6}(\beta\mathcal{H}\mu)^2 \right] \quad (17.51)$$

---

† We include a factor of $1/4\pi$ in Eq.(17.48) so that $z_{\text{int}}(\mathcal{H}) \longrightarrow 1$ as $\mathcal{H} \longrightarrow 0$, and hence $z_{\text{int}} \equiv (z_{\text{int}})_0 z_{\text{int}}(\mathcal{H})$. It is evident from Eqs.(17.54-17.56) that this factor has no effect on the magnetization.

The logarithm of the partition function is given by

$$ln\mathcal{Z}_{\text{int}}(\mathcal{H}) = Nlnz_{\text{int}}(\mathcal{H}) \tag{17.52}$$

In weak fields the logarithm of the second factor in Eq.(17.51) can be expanded as $ln(1+x) \cong x$ and hence

$$ln\mathcal{Z}_{\text{int}}(\mathcal{H}) \cong N \left[\frac{(\beta\mu\mathcal{H})^2}{6} - \frac{\beta Ze^2r^2\mathcal{H}^2}{12mc^2} \right] \tag{17.53}$$

That part of the free energy which depends on $\mathcal{H}$ is then obtained from Eq.(11.5) as

$$F(\mathcal{H}) = -\frac{1}{\beta}ln\mathcal{Z}_{\text{int}}(\mathcal{H}) \tag{17.54}$$

and hence

$$F(\mathcal{H}) \cong N \left[\frac{-\beta\mu^2\mathcal{H}^2}{6} + \frac{Ze^2r^2\mathcal{H}^2}{12mc^2} \right] \tag{17.55}$$

There is now a magnetic field dependence in this expression as advertised, and the induced magnetization can be obtained from Eq.(17.13) as

$$M = -\left(\frac{\partial F}{\partial \mathcal{H}}\right)_{V,T} \cong \frac{\beta N\mu^2\mathcal{H}}{3} - \frac{NZe^2r^2\mathcal{H}}{6mc^2} \tag{17.56}$$

Hence we have derived the important result that

$$M = \left(\frac{\beta N\mu^2}{3} - \frac{NZe^2r^2}{6mc^2}\right)\mathcal{H} \qquad ; \text{ weak fields} \tag{17.57}$$

The induced magnetization is linear in the field strength $\mathcal{H}$. The magnetic dipole moment per unit volume yields the magnetic susceptibility through Eq.(17.2)

$$\chi = \frac{n\mu^2}{3k_BT} - \frac{nZe^2r^2}{6mc^2} \tag{17.58}$$

where n is the density of atoms

$$n \equiv N/V \tag{17.59}$$

This result is known as the *Langevin-Curie theory of magnetism*.

The first term in Eq.(17.58) is the atomic contribution to *paramagnetism*. It vanishes if the atomic magnetic moment μ is zero. The paramagnetic susceptibility is inversely proportional to the temperature T; this is known as the *Curie law*. The second term in Eq.(17.58), which enters with opposite sign, contributes to atomic *diamagnetism*. It is always present, since all atoms have a finite size. The diamagnetic contribution is independent of the temperature T, and is the entire contribution to the magnetic susceptibility for substances such as the noble gases for which $\mu = 0$. Note that this theory, developed before quantum mechanics, was able to describe both diamagnetism and paramagnetism (albeit with *ad hoc* assumptions).

For orientation, we estimate the order of magnitude of the two contributions in Eq.(17.58). The following relations on fundamental constants are useful here

$$\frac{e^2}{m_e c^2} = 2.819 \times 10^{-13} \text{cm}$$

$$k_B = 1.381 \times 10^{-16} \text{erg}/^\circ K \qquad (17.60)$$

$$1 \text{ erg} \equiv 1 \text{ Gauss}^2 \text{cm}^3$$

Typical physical values appearing in Eq.(17.58) are

$$n \sim 10^{23}/\text{cm}^3 \qquad \mu \sim 10^{-20} \text{erg}/\text{Gauss}$$

$$T \sim 300^\circ K \qquad Zr^2 \sim 10^{-16} \text{cm}^2 \qquad (17.61)$$

These values imply that

$$\chi \text{para} \equiv \frac{n\mu^2}{3k_B T} \sim 10^{-4} \qquad (17.62)$$

and

$$\chi \text{dia} \equiv \frac{nZe^2 r^2}{6m_e c^2} \sim 10^{-6} \qquad (17.63)$$

Note that the susceptibility is *dimensionless*.

It is possible to evaluate the integral for the partition function analytically for arbitrary field strengths in Eq.(17.48). Define the expression in Eq.(17.48) to be

$$z_{\text{int}}(\mathcal{H}) = z_{\text{dia}}(\mathcal{H}) z_{\text{para}}(\mathcal{H}) \qquad (17.64)$$

where

$$z_{\text{para}}(\mathcal{H}) \equiv \frac{1}{4\pi} \int_0^{2\pi} d\phi \int_0^\pi \sin\theta d\theta \, e^{\beta\mu\mathcal{H}\cos\theta} \qquad (17.65)$$

Then a simple change of variables $x = \cos\theta$ converts the paramagnetic contribution to the partition function to

$$z_{\text{para}}(\mathcal{H}) = \frac{1}{2} \int_{-1}^1 dx \, e^{\beta\mu\mathcal{H}x} \qquad (17.66)$$

with the result that

$$z_{\text{para}}(\mathcal{H}) = \frac{1}{2\beta\mu\mathcal{H}} \left(e^{\beta\mu\mathcal{H}} - e^{-\beta\mu\mathcal{H}} \right) \qquad (17.67)$$

The corresponding free energy and magnetization are readily evaluated. At very high fields where $\mu\mathcal{H}/k_B T >> 1$, the results are particularly simple

$$z_{\text{para}}(\mathcal{H}) \cong \frac{1}{2\beta\mu\mathcal{H}} e^{\beta\mu\mathcal{H}} \qquad (17.68)$$

$$F_{\text{para}}(\mathcal{H}) \equiv -\frac{N}{\beta} ln z_{\text{para}}(\mathcal{H}) \cong -\frac{N}{\beta}[\beta\mu\mathcal{H} - ln(2\beta\mu\mathcal{H})] \, (17.69)$$

$$M_{\text{para}}(\mathcal{H}) = -\left(\frac{\partial F_{\text{para}}(\mathcal{H})}{\partial \mathcal{H}} \right)_{VT} \cong N\mu \left(1 - \frac{1}{\beta\mu\mathcal{H}} \right)$$

$$M_{\text{para}}(\mathcal{H})/V \cong n\mu(1 - k_B T/\mu\mathcal{H}) \quad ; \text{ high fields} \qquad (17.70)$$

The dependence of the magnetization on the applied field in this Langevin-Curie theory is sketched in Figure 17.4. It exhibits the previously discussed linear dependence at low fields, where the tendency to lower the energy and align the magnetic moments of the subsystems along the field is counterbalanced by the thermal randomness of the canonical ensemble. At high fields there is *saturation* which arises because the applied field can never do more than align all the magnetic moments along the field. The saturation value of the magnetization is evidently $M/V = n\mu$.

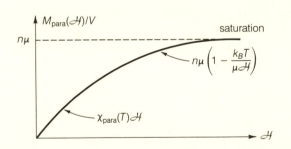

Figure 17.4 Sketch of the magnetization $M_{\text{para}}(H)/V$ in the Langevin-Curie theory of magnetism.

VI QUANTUM STATISTICS

Our discussion of the statistical ensemble, thermal equilibrium, and the canonical distribution so far has been based on classical mechanics. In this description, the state of a many-particle system is completely specified, in principle, by giving experimental values to the generalized coordinates and momenta $(p_i, q_i; \; i = 1, 2, \ldots, f)$, where f is the total number of degrees of freedom.† The time development of the system is then prescribed by Hamilton's equations. The essential difference between quantum mechanics and classical mechanics is that in quantum mechanics it is no longer possible to simultaneously know both q_i and p_i, since measurement of q_i affects the knowledge of p_i and vice versa. There is an uncertainty Δp_i and Δq_i in the simultaneous specification of the values of $(p_i, q_i; \; i = 1, 2, \ldots, f)$ given by the *Heisenberg uncertainty principle*.

$$\Delta p_i \Delta q_i \sim \hbar \qquad ; \; i = 1, 2, \ldots, f \qquad (18.0)$$

Our discussion of statistical mechanics must therefore be modified in the quantum domain. To prepare for this modification, we review the essential elements of quantum mechanics.

18. Basic Elements of Quantum Mechanics

In quantum mechanics the state of the system, rather than being specified by $(p_i, q_i; \; i = 1, 2, \ldots, f)$, is given by a *wavefunction*

$$\psi(q_j) \equiv \psi(q_1, q_2, \ldots, q_f) \qquad (18.1)$$

where the *probability* of finding the coordinates q_j between q_j and $q_j + dq_j$ for $j = 1, 2, \ldots, f$ is given by

$$\psi^*(q_j)\psi(q_j) \prod_{j=1}^{f} dq_j = \text{probability of finding } q_j \text{ between } q_j \text{ and}$$

$$q_j + dq_j \text{ for } j = 1, 2, \ldots, f$$

$$(18.2)$$

We also have

$$\int |\psi(q_j)|^2 \prod_{j=1}^{f} dq_j = 1 \qquad (18.3)$$

---

† Note that we henceforth denote the total number of degrees of freedom of the system by f.

We also have

$$\int |\psi(q_j)|^2 \prod_{j=1}^{f} dq_j = 1 \tag{18.3}$$

One can equivalently go to momentum space by taking the Fourier transform

$$\psi(p_j) \equiv \left(\frac{1}{\sqrt{h}}\right)^f \int e^{i/\hbar \sum_{j=1}^{f} p_j q_j} \, \psi(q_j) \prod_{j=1}^{f} dq_j \tag{18.4}$$

and the wavefunction in momentum space has the similar interpretation that the probability to find the momenta p_j with values between p_j and $p_j + dp_j$ is given by

$$\psi^*(p_j)\psi(p_j) \prod_{j=1}^{f} dp_j = \text{probability of finding } p_j \text{ between } p_j \text{ and}$$

$$p_j + dp_j \text{ for } j = 1, 2, \ldots, f$$
$$\tag{18.5}$$

It follows from the completeness of the Fourier transform in Eq.(18.4) that†

$$\int \psi^*(p_j)\psi(p_j) \prod_j dp_j = \int \psi^*(q_j)\psi(q_j) \prod_j dq_j = 1 \tag{18.6}$$

For example, if the coordinates have gaussian distributions about the values $(q_j'; \; j = 1, 2, \ldots, f)$

$$\psi(q_j) \sim \prod_j e^{-\alpha^2(q_j - q_j')^2/2} \tag{18.7}$$

then an elementary calculation shows that the momenta $(p_j; \; j = 1, 2, \ldots, f)$ will also have gaussian distributions

$$\psi(p_j) \sim \prod_j e^{-p_j^2/2\hbar^2\alpha^2} \tag{18.8}$$

The situation is illustrated in Figure 18.1. The half-widths of the probability distributions obtained from Eqs.(18.7-18.8) and illustrated in Figure 18.1 are evidently given by

$$\Delta q_j \sim \frac{1}{\alpha} \tag{18.9}$$
$$\Delta p_j \sim \alpha\hbar$$

---

† The volume elements will henceforth be denoted by $\prod_j dq_j$ and $\prod_j dp_j$.

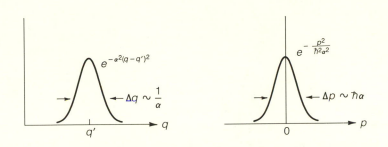

Figure 18.1 Example of probability distributions obtained from coordinate space wavefunction with gaussian distribution of the coordinate q about q' [Eq.(18.7)] and corresponding momentum space wavefunction obtained by taking the Fourier transform [Eq.(18.8)].

These half-widths provide a measure of the uncertainty of the values of the corresponding quantities, and the product of the two values in Eqs.(18.9) provides an explicit illustration of the Heisenberg uncertainty principle

$$\Delta p_j \Delta q_j \sim \hbar \qquad (18.10)$$

Now consider a dynamical quantity $Q = Q(p_j, q_j)$. In quantum mechanics one must replace this by an *operator*. In coordinate space we write

$$Q(p_j, q_j) \longrightarrow Q\left(\frac{\hbar}{i}\frac{\partial}{\partial q_j}, q_j\right) \qquad (18.11)$$

The *"expectation value"* $< Q >$ of the dynamical quantity Q in the state ψ is then given by the expression

$$< Q > \equiv \int \psi^*(q_j) Q\left(\frac{\hbar}{i}\frac{\partial}{\partial q_j}, q_j\right)\psi(q_j)\prod_j dq_j \qquad (18.12)$$

Given the wavefunction $\psi(q_j, t_0)$ at a time t_0, one obtains the wavefunction $\psi(q_j, t)$ at a later time t from the time-dependent *Schrödinger equation*

$$-\frac{\hbar}{i}\frac{\partial\psi}{\partial t} = H\psi \qquad (18.13)$$

The hamiltonian appearing in this expression is the operator obtained from $H(p_j, q_j)$ with the replacement of Eq.(18.11)

$$H(p_j, q_j) \longrightarrow H\left(\frac{\hbar}{i}\frac{\partial}{\partial q_j}, q_j\right) \qquad (18.14)$$

In this *Schrödinger picture* of quantum mechanics, the expectation value of the dynamical quantity $Q(p_j, q_j)$ changes with time due to the time development of the wavefunction $\psi(q_j, t)$ as specified by the Schrödinger equation in Eq.(18.13).

If the hamiltonian in Eq.(18.14) indeed contains no *explicit* time dependence, one may separate variables in the time-dependent Schrödinger equation and look for solutions of the form

$$\psi(q_j, t) = \psi(q_j)e^{-iEt/\hbar} \qquad (18.15)$$

Substitution in Eq.(18.13) and cancellation of common factors then leads to the time-independent Schrödinger equation

$$H\left(\frac{\hbar}{i}\frac{\partial}{\partial q_j}, q_j\right)\psi(q_j) = E\psi(q_j) \qquad (18.16)$$

Under the imposition of appropriate *boundary conditions* this differential equation will possess a sequence of eigenvalues and eigenfunctions

$$H\phi_n = E_n\phi_n \qquad ; \; n = 1, 2, \ldots, \infty \qquad (18.17)$$

It is evident from Eq.(18.12) that the expectation value of the dynamical quantity $Q(p_j, q_j)$ will be independent of time when evaluated with a wavefunction of the form in Eq.(18.15); these wavefunctions are said to correspond to *stationary states*. We can take the set of eigenfunctions in Eq.(18.17) to be orthonormal

$$\int \phi_m^*(q_j)\phi_n(q_j) \prod_j dq_j = \delta_{mn} \qquad (18.18)$$

The eigenfunctions are also assumed to be complete. This implies that at any instant t, an arbitrary wavefunction can be expanded as

$$\psi(q_j, t) = \sum_n a_n(t)\phi_n(q_j) \qquad (18.19)$$

where the coefficients $a_n(t)$ will depend on the time. The expectation value of $Q(p_j, q_j)$ at the time t in the state described by the wavefunction in Eq.(18.19) is given by

$$
\begin{aligned}
< Q(t) > &= \sum_m \sum_n a_m^*(t) a_n(t) \int \phi_m^*(q_j) Q\left(\frac{\hbar}{i}\frac{\partial}{\partial q_j}, q_j\right) \phi_n(q_j) \prod_j dq_j \\
&= \sum_m \sum_n a_m^*(t) a_n(t) Q_{mn}
\end{aligned}
$$

(18.20)

where the matrix elements Q_{mn} of the operator $Q(p_j, q_j)$ are defined by

$$
Q_{mn} \equiv \int \phi_m^*(q_j) Q\left(\frac{\hbar}{i}\frac{\partial}{\partial q_j}, q_j\right) \phi_n(q_j) \prod_j dq_j \qquad (18.21)
$$

The expectation value in Eq.(18.20) changes with time because the coefficients $a_n(t)$ depend on time; their time development is determined by substituting the wavefunction in Eq.(18.19) into the time-dependent Schrödinger Eq.(18.13)

$$
-\frac{\hbar}{i} \sum_n \dot{a}_n(t) \phi_n(q_j) = \sum_n a_n(t) H \phi_n(q_j) \qquad (18.22)
$$

The orthonormality of the eigenfunctions in Eq.(18.18) permits us to write

$$
-\frac{\hbar}{i}\dot{a}_n(t) = \sum_m H_{nm} a_m(t) \qquad (18.23)
$$

where the matrix elements of the hamiltonian are defined by

$$
H_{mn} \equiv \int \phi_m^*(q_j) H\left(\frac{\hbar}{i}\frac{\partial}{\partial q_j}, q_j\right) \phi_n(a_j) \prod_j dq_j \qquad (18.24)
$$

It will be assumed that the boundary conditions of the problem are such that when partial integration is performed on the derivatives appearing in the operator in Eq.(18.21), the boundary contributions disappear. As a consequence, the expectation value of the operator in Eq.(18.12) is real

$$
< Q > = \int \left[Q\left(\frac{\hbar}{i}\frac{\partial}{\partial q_j}, q_j\right) \psi(q_j) \right]^* \psi(q_j) \prod_j dq_j = < Q >^* \qquad (18.25)
$$

and the matrix elements of this operator in Eq.(18.21) are *hermitian*

$$
Q_{mn}^* = Q_{nm} \qquad (18.26)
$$

For an operator to correspond to a physical observable, its expectation value must be real. Physical observables in quantum mechanics are thus represented by such *hermitian* operators. In particular, the matrix elements of the hamiltonian must be hermitian

$$H_{mn}^* = H_{nm} \tag{18.27}$$

The complex conjugate of Eq.(18.23) then leads to the following equation for the time development of $a_n^*(t)$

$$\frac{\hbar}{i} \dot{a}_n^*(t) = \sum_m a_m^*(t) H_{mn} \tag{18.28}$$

Furthermore, substitution of the expansion in Eq.(18.19) into the normalization condition of Eq.(18.3) yields

$$\int \psi^*(q_j)\psi(q_j) \prod_j dq_j = \sum_{mn} a_m^* a_n \int \phi_m^*(q_j)\phi_n(q_j) \prod_j dq_j = 1 \tag{18.29}$$

Use of the orthonormality condition of Eq.(18.18) leads to the following relation, which must be satisfied by the expansion coefficients

$$\sum_m \left| a_m(t) \right|^2 = 1 \tag{18.30}$$

19. The Density Matrix

We now introduce the *density matrix* defined by

$$\rho_{nm} = \rho_{nm}(t) \equiv a_m^*(t) a_n(t) \quad ; \text{ density matrix} \tag{19.1}$$

It is a matrix constructed from the expansion coefficients in the wave function in Eq.(18.19). The interpretation of these coefficients in quantum mechanics is that their absolute square $|a_n(t)|^2$ gives the probability of finding the system in the state n at time t. The diagonal elements of the density matrix are just these probabilities

$$\rho_{nn} = |a_n(t)|^2 = \text{probability of finding the system in the state } n$$
$$\text{at the time } t$$

$$\tag{19.2}$$

The normalization condition in Eq.(18.30) states that the sum of the probabilities must be unity

$$\sum_n \rho_{nn} = 1 \qquad (19.3)$$

Since the sum of the diagonal elements is just the trace of the matrix, this relation can be written in the more compact form

$$\text{Trace } \underline{\rho} = 1 \qquad (19.4)$$

The expectation value of the hermitian operator which represents the dynamical quantity Q in Eq.(18.20) can be written in terms of the density matrix as

$$< Q >= \sum_m \sum_n Q_{mn}\rho_{nm} \qquad (19.5)$$

Identification of the matrix product permits this to be rewritten as

$$< Q >= \sum_m (\underline{Q}\underline{\rho})_{mm} \qquad (19.6)$$

or

$$< Q >= \text{Trace } (\underline{Q}\underline{\rho}) \qquad (19.7)$$

The matrix sum runs over the complete set of eigenstates, and one must know the whole matrix $\underline{\rho}$ to evaluate $< Q >$.

The density matrix defined in Eq.(19.1) is a function of time, because the expansion coefficients in the wavefunction in Eq.(18.19) are time-dependent; their time development is given by the Schrödinger equation. Thus we can compute

$$-\frac{\hbar}{i}\frac{d\rho_{nm}(t)}{dt} = -\frac{\hbar}{i}\left[\left(\frac{da_m^*}{dt}\right)a_n + a_m^*\left(\frac{da_n}{dt}\right)\right] \qquad (19.8)$$

and substitution of the equations of motion in Eqs.(18.23) and (18.28) gives

$$-\frac{\hbar}{i}\frac{d\rho_{nm}(t)}{dt} = \sum_p \left(a_m^* H_{np}a_p - a_p^* H_{pm}a_n\right) \qquad (19.9)$$

The identification of the density matrix permits this expression to be rewritten as

$$-\frac{\hbar}{i}\frac{d\rho_{nm}(t)}{dt} = \sum_p \left(H_{np}\rho_{pm} - \rho_{np}H_{pm}\right) \qquad (19.10)$$

or, in matrix notation,

$$-\frac{\hbar}{i}\frac{d\rho_{nm}(t)}{dt} = (\underline{H}\rho)_{nm} - (\rho\underline{H})_{nm} \qquad (19.11)$$

This expression is just the difference in the matrix products $\underline{H}\rho$ and $\rho\underline{H}$, which defines the commutator of these two matrices. Thus one has finally

$$-\frac{\hbar}{i}\frac{d\rho_{nm}}{dt} = [\underline{H}, \rho]_{nm} \qquad (19.12)$$

This expression is the main result of this section. It is the quantum analogue of the classical relations in Eqs.(6.35) and (6.36) for the probability density ρ. In words it states that the time development of the density matrix, from which the expectation value of any hermitian operator can be calculated by Eq.(19.7), is given by the commutator with the hamiltonian matrix of Eq.(18.24). Since the trace is invariant under cyclic permutations, it follows from Eq.(19.12) that

$$\sum_n \frac{d}{dt}\rho_{nn} = \frac{d}{dt}\text{Trace } \underline{\rho} = 0 \qquad (19.13)$$

Thus Trace $\underline{\rho}$ is a constant of the motion, and it must satisfy the normalization condition of Eq.(19.4). Equations (19.12), (19.13), and (19.7) form the basis for our discussion of quantum statistical mechanics. We proceed to an analysis of the quantum statistical ensemble.

20. The Statistical Ensemble

The state of a classical system is completely determined by giving values to the quantities $(p_i, q_i; \; i = 1, 2, \ldots, f)$, where f is the number of degrees of freedom. In quantum mechanics, the values of p_i and q_i cannot be simultaneously specified due to the uncertainty principle. How, then, do we completely specify the corresponding quantum state? As a start, the stationary state defined by Eqs.(18.15-16) has a well-defined energy. This is easily demonstrated. It follows from Eq.(18.16) that the expectation value of the hamiltonian in this state is just the constant E

$$< H > = \int \psi^*(q_j)e^{iEt/\hbar} H\left(\frac{\hbar}{i}\frac{\partial}{\partial q_j}, q_j\right)\psi(q_j)e^{-iEt/\hbar}\prod_j dq_j = E \quad (20.1)$$

Furthermore, since a stationary state is an eigenstate of the hamiltonian, all moments of the hamiltonian about E must vanish

$$< (H - E)^n > = 0 \qquad ; \qquad \text{all } n \qquad (20.2)$$

The energy thus indeed has the precise value E in a stationary state. But to completely determine the quantum state of a system, many other quantities must be specified.

The set of eigenfunctions in Eq.(18.17) can be defined by stating that a "complete set" of observables O_k with $k = 1, 2, \ldots, f$ has the property that

$$O_k \phi_n = (O_k')_n \phi_n \quad ; \; k = 1, 2, \ldots, f$$
$$n = 1, 2, \ldots, \infty$$

(20.3)

The number of members of the set of observables O_k with $k = 1, 2, \ldots, f$ must be equal to the total number of degrees of freedom. Equation (20.3) states that in the state ϕ_n of the system, the observable O_k will be found to have the eigenvalue $(O_k')_n$. Then in this "O_k-representation" the matrix elements of the operators O_k will have the form

$$(O_k)_{mn} = (O_k')_n \delta_{mn} \tag{20.4}$$

They form a diagonal matrix

$$\underline{O}_k = \begin{array}{c} m\backslash n \\ 1 \\ 2 \\ 3 \end{array} \begin{array}{cccc} 1 & 2 & 3 & \\ \left(\begin{matrix} (O_k')_1 & & & \\ & (O_k')_2 & & \\ & & (O_k')_3 & \\ & & & \ddots \end{matrix} \right) \end{array} \tag{20.5}$$

Now the operators O_k with $k = 1, 2, \ldots, f$ can only have the common eigenstates of Eq.(20.3) if they *mutually commute*. This is readily demonstrated. Apply the operators first in one order

$$O_k O_l \phi_n = O_k (O_l')_n \phi_n = (O_k')_n (O_l')_n \phi_n \tag{20.6}$$

and then in the other

$$O_l O_k \phi_n = (O_l')_n (O_k')_n \phi_n \equiv (O_k')_n (O_l')_n \phi_n \tag{20.7}$$

It follows that

$$(O_k O_l - O_l O_k)\phi_n \equiv [O_k, O_l]\phi_n = 0 \tag{20.8}$$

Upon performing the operation $O_k O_l - O_l O_k$, defined as the commutator $[O_k, O_l]$, one gets zero; it follows that all matrix elements of the commutator $[O_k, O_l]$ vanish

$$([O_k, O_l])_{mn} = 0 \quad ; \; \text{all } m, n \tag{20.9}$$

Since the basis of eigenstates is assumed to be complete, this implies that the operator itself must vanish

$$[O_k, O_l] = 0 \tag{20.10}$$

as claimed.

We shall take the hamiltonian to be the first member of the complete set O_k

$$O_1 \equiv H \tag{20.11}$$

Thus

$$H\phi_n = E_n\phi_n \tag{20.12}$$

The matrix elements of this operator are then given by

$$H_{mn} = E_n\delta_{mn} \tag{20.13}$$

and this matrix is indeed diagonal as in Eq.(20.14). The diagonal elements are just the energy eigenvalues

$$\underline{O_1} \equiv \underline{H} = \begin{pmatrix} E_1 & & & \\ & E_2 & & \\ & & E_3 & \\ & & & \ddots \end{pmatrix} \tag{20.14}$$

The same eigenvalue may appear several times if there is a *degeneracy*. We must now further specify the eigenvalues of the remaining members of the complete set of observables $O_2, O_3, \ldots, O_f$, all of which must commute with the hamiltonian, which is the first member of the set.

The hydrogen atom provides a familiar example. In the absence of spin, there are just three degrees of freedom. In addition to H, one may choose as observables the z-component of the angular momentum and the square of the total angular momentum†

$$O_2 \equiv L_z = \frac{1}{\hbar}(\mathbf{r} \wedge \mathbf{p})_z \tag{20.15}$$

$$O_3 \equiv \mathbf{L}^2 = L_x^2 + L_y^2 + L_z^2 \tag{20.16}$$

The eigenvalues of these operators are

$$L_z\phi_{\bar{n}lm_l} = m_l\phi_{\bar{n}lm_l} \quad ; \ -l \leq m_l \leq l \tag{20.17}$$

$$\mathbf{L}^2\phi_{\bar{n}lm_l} = l(l+1)\phi_{\bar{n}lm_l} \quad ; \ l = 0, 1, 2, \ldots, \infty \tag{20.18}$$

---

† Note that here and henceforth we define the angular momentum in units of $\hbar$.

In this case, the state of the system (the hydrogen atom) can be characterized by the set of quantum numbers $n \longrightarrow (\bar{n}, l, m_l)$, where the remaining $\bar{n}$ is now the "principal quantum number." The familiar spectrum of the hydrogen atom is sketched in Figure 20.1.

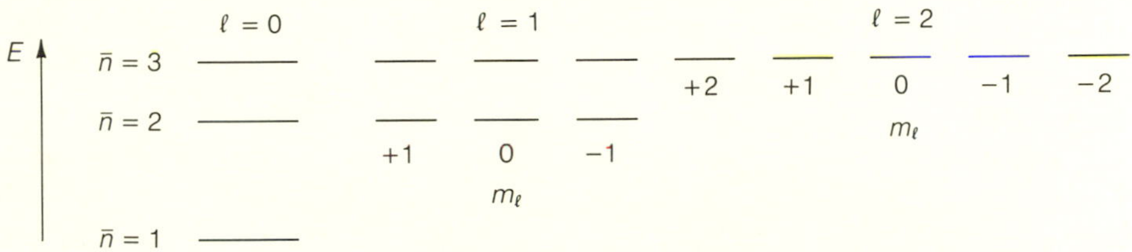

Figure 20.1 Low-lying spectrum of the non-relativistic, spinless hydrogen atom in quantum mechanics. The complete set of states at each energy is indicated. The quantum numbers are $(\bar{n}, l, m_l)$, where $\bar{n}$ is the principal quantum number; the label n here denotes an ordered counting of these states.

In the case of the hydrogen atom, there are $\bar{n}^2$ states at each energy; there is an $\bar{n}^2$-fold "degeneracy." The degeneracy of the states with different l in this system arises because of the *symmetry* of the $1/r$ Coulomb potential.

If all observables O_k have been measured at a certain time and found to have the values $(O'_k)_n$, one knows with certainty that the system is in the state ϕ_n. In this case the coefficients a_n in Eq.(18.19) must have the form

$$|a_n| = 1$$
$$a_p = 0 \text{ for } p \neq n \tag{20.19}$$

Otherwise, one can only say that the system is in the state

$$\psi = \sum_n a_n \phi_n \tag{20.20}$$

with

$$\sum_n |a_n|^2 = 1 \tag{20.21}$$

One can then only make *probability* statements about the individual a_n. The *statistical* element emerges in figuring out how the a_n will be distributed.

Consider, then, an *ensemble* of systems. Since the coefficients a_n are complex, we write

$$a_n = \alpha_n + i\beta_n \qquad (20.22)$$

The normalization condition of Eq.(20.21) then becomes

$$\sum_n (\alpha_n^2 + \beta_n^2) = 1 \qquad (20.23)$$

To completely specify the state of a given system one must specify

$$\text{Specify } (\alpha_1, \beta_1, \alpha_2, \beta_2, \ldots \alpha_p, \beta_p, \ldots) \qquad (20.24)$$

It is convenient to relabel these parameters as an ordered set of real quantities $(\xi_n;\ n = 1, 2, \ldots, 2p, \ldots, \infty)$

$$\text{Relabel } (\xi_1, \xi_2, \ldots, \xi_{2p}, \ldots) \equiv (\xi_n) \qquad (20.25)$$

The number of members of the ensemble that have values lying between ξ_n and $\xi_n + d\xi_n$ for $n = 1, 2, \ldots, \infty$ will now be denoted by

$$
\begin{aligned}
d\nu =& D(\xi_n) \prod_n d\xi_n \\
=& \text{Number of members of the ensemble with} \\
& \xi_n \text{ between } \xi_n \text{ and } \xi_n + d\xi_n \text{ for } n = 1, 2, \ldots, \infty
\end{aligned}
\qquad (20.26)
$$

The *distribution function* $D(\xi_n)$ in parameter space now plays a role analogous to that of the classical distribution function in phase space [Eq.(4.4)]. In the quantum case the *quantum phase space* (or ξ_n-space) is characterized by assigning values to the real parameters $(\xi_n;\ n = 1, 2, \ldots, \infty)$; it is infinite-dimensional. The normalization condition of Eq.(20.23)

$$\sum_n \xi_n^2 = 1 \qquad (20.27)$$

restricts us to the unit sphere in this space. In the classical case, the distribution is specified in phase space $(p_i, q_i;\ i = 1, 2, \ldots, f)$, which is finite-dimensional, but all values of the momenta and coordinates must be considered.

In direct analogy to the classical case, the *quantum statistical ensemble* will be specified by assigning systems to a set of points in the quantum phase space. In general, just as in Section 6, the distribution function will have an explicit dependence on the time due to the difference of the number of points entering and leaving a given volume element and we will write $D(\xi_n, t)$.

Given the coefficients $a_n(t_0)$ at the time t_0, their value $a_n(t)$ at the subsequent time t is obtained by forward integration of the Schrödinger Eq.(18.23). Thus, just as in Eq.(3.2), one can write the functional relationship

$$\xi'_p = \xi'_p(\xi_n) \qquad ; \; p = 1, 2, \ldots, \infty \tag{20.28}$$

Define the volume elements as

$$d\lambda_0 \equiv \prod_{n=1}^{\infty} d\xi_n$$
$$d\lambda_t \equiv \prod_{n=1}^{\infty} d\xi'_n \tag{20.29}$$

The quantum analog of Liouville's theorem then states that the volume element in the quantum phase space is unchanged along a phase orbit

$$d\lambda_t = d\lambda_0 \qquad ; \; \text{along a phase orbit} \tag{20.30}$$

The proof follows as in Section 3 (Problem 20.1). In fact, this result is readily obtained in the following simpler fashion. Equation (19.13) establishes the result that the normalization condition of Eq.(18.30) is preserved in time by the Schrödinger equation; this is the statement of *unitarity*. In ξ-space one thus has

$$\sum_n \xi'^2_n = \sum_n \xi^2_n \tag{20.31}$$

In ξ-space the Schrödinger equation therefore gives rise to a *real rotation* that leaves the point on the unit sphere. Since a real rotation preserves volume elements, the result in Eq.(20.30) follows immediately. The quantum statistical ensemble is indicated schematically in Figure 20.2, which is the quantum analog of Figure 6.1.

It is convenient to include the restriction to the unit sphere analytically in the volume element in the quantum phase space, and we henceforth define

$$d\lambda \equiv \delta\left(1 - \sum_{n=1}^{\infty} \xi^2_n\right) \prod_{n=1}^{\infty} d\xi_n \tag{20.32}$$

Since the Schrödinger equation preserves the norm in Eq.(20.31), the quantum version of Liouville's theorem in Eq.(20.30) still holds.

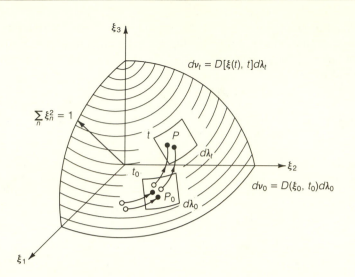

Figure 20.2 Schematic representation of the time evolution of the representative points in the quantum phase space of the members of the quantum statistical ensemble. Here (ξ) stands for the variable set (ξ_n) or $(\alpha_n, \beta_n;\ n = 1, 2, \ldots, \infty)$, where the expansion coefficients in Eq.(18.19) are written as $a_n = \alpha_n + i\beta_n$. The volume elements of Eq.(20.32) are indicated.

The quantum distribution function will henceforth be similarly defined according to

$$d\nu \equiv D(\xi, t)d\lambda$$

$$= \text{Number of members of the ensemble with values of} \qquad (20.33)$$

$$\xi_n \text{ between } \xi_n \text{ and } \xi_n + d\xi_n \text{ for } n = 1, 2, \ldots, \infty$$

where (ξ) now stands for the variable set $(\xi_n;\ n = 1, 2, \ldots, \infty)$.

Once the quantum distribution function has been specified at some initial time t_0, its time development follows from the general principles of quantum mechanics. Let the number of members of the ensemble $d\nu_0$ within the volume element $d\lambda_0$ at the time t_0 be denoted by

$$d\nu_0 = D(\xi_0, t_0)d\lambda_0 \qquad (20.34)$$

Then as one proceeds along the phase orbit to the time t

$$d\nu_t = D[\xi(t), t]d\lambda_t \qquad (20.35)$$

But as the number of members of the ensemble does not change

$$dv_0 = dv_t \qquad (20.36)$$

nor does the volume element

$$d\lambda_0 = d\lambda_t \qquad (20.37)$$

we have

$$D[\xi(t), t] = D(\xi_0, t_0) = \text{const.} \qquad (20.38)$$

Hence, once again, *the distribution function is unchanged along a phase orbit*. Differentiation of this relation with respect to time yields the relation

$$\frac{dD}{dt} = \sum_n \frac{\partial D}{\partial \xi_n} \dot{\xi}_n + \frac{\partial D}{\partial t} = 0 \qquad (20.39)$$

This is the differential equation which the quantum distribution function must satisfy. It is known as Ehrenfest's theorem and forms the quantum analog of Liouville's theorem as discussed in Section 6.

Now knowledge of the density matrix ρ in Eq.(19.1) provides a complete description of any quantum system. Since one is never in a position to make a complete set of measurements on a many-particle system with a vast number of degrees of freedom, it is only the *statistical average density matrix* which determines physical observables. This statistical average is defined by taking the average over the quantum statistical ensemble using the distribution in Eq.(20.33). For any element ρ_{nm} of the density matrix ρ we must take the average over the ensemble

$$\bar{\rho}_{nm} = \overline{a_m^* a_n} \equiv \frac{1}{\nu} \int a_m^* a_n D(\xi, t) d\lambda$$

$$\equiv \frac{1}{\nu} \int a_m^* a_n D(\xi, t) \delta(1 - \sum_n \xi_n^2) \prod_n d\xi_n \qquad (20.40)$$

; statistical density matrix

where the bar over the symbol now indicates the *statistical density matrix*. Here ν denotes the total number of systems defined by

$$\nu \equiv \int D(\xi, t) d\lambda$$

$$= \int D(\xi, t) \delta(1 - \sum_n \xi_n^2) \prod_n d\xi_n \qquad (20.41)$$

The statistical average of any observable is now obtained directly by taking the statistical average of (19.5-19.7)

$$\overline{< Q(t) >} = \sum_{mn} Q_{mn} \bar{\rho}_{nm}$$

$$= \text{Trace } (Q\underline{\bar{\rho}})$$

(20.42)

It is important to note that the statistical density matrix can be quite different from the actual one. For example, suppose that the statistical density matrix is diagonal

$$\bar{\rho}_{nm} = \bar{\rho}_n \delta_{mn}$$

(20.43)

as will turn out to be the case in our subsequent discussion. Now consider an individual system and assume, without loss of generality, that there is one non-zero coefficient $a_n \neq 0$ in the wavefunction. If applied to the *true* density matrix for this system in Eq.(19.1), the diagonal condition would imply that all a_m with $m \neq n$ must vanish. It follows that $a_m^* a_m = 0$ for all $m \neq n$ and hence if the true density matrix is to be diagonal, it can have only one non-zero element $a_n^* a_n$! In short, the only way the true density matrix ρ can be diagonal is if one diagonal element is unity; this implies certain knowledge that the system is in one particular state!

Before proceeding, it is convenient to introduce polar coordinates in ξ-space. Define

$$a_n = \alpha_n + i\beta_n \equiv |a_n|e^{i\phi_n}$$

(20.44)

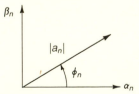

Figure 20.3 Polar coordinates in ξ-space. Recall $(\alpha_n, \beta_n) = (\xi_{2n-1}, \xi_{2n})$ (Eqs.20.24-20.25)

as illustrated in Figure 20.3. It follows that

$$\prod_n d\xi_n = \prod_n |a_n|d|a_n|d\phi_n \tag{20.45}$$

and also

$$\sum_n \xi_n^2 = \sum_n |a_n|^2 \tag{20.46}$$

The volume element in quantum phase space in Eq.(20.32) then takes the form

$$d\lambda = \delta(1 - \sum_n |a_n|^2)\prod_n |a_n|d|a_n|d\phi_n \tag{20.47}$$

The statistical density matrix of Eq.(20.40) can thus be written as

$$\bar{\rho}_{nm} = \frac{1}{\nu}\int a_m^* a_n D(|a_n|,\phi_n;t)\delta(1 - \sum_n |a_n|^2)\prod_n |a_n|d|a_n|d\phi_n \tag{20.48}$$

This can be rewritten with the aid of Eq.(20.44) as

$$\bar{\rho}_{nm} = \frac{1}{\nu}\int |a_m||a_n|e^{i(\phi_n-\phi_m)}D(|a_n|,\phi_n;t)\delta(1 - \sum_n |a_n|^2)\prod_n |a_n|d|a_n|d\phi_n \tag{20.49}$$

Now suppose that the distribution function is independent of the detailed phases in the wave function so that

$$D(|a_n|,\phi_n;t) = D(|a_n|,t) \tag{20.50}$$

This will, in fact, turn out to be the case in the following discussion. Since

$$\int_0^{2\pi} e^{i(\phi_n-\phi_m)}d\phi_n d\phi_m = 0 \quad \text{for } n \neq m \tag{20.51}$$

It then follows from Eq. (20.49) that the statistical density matrix will indeed be *diagonal* as in Eq.(20.43) with diagonal elements given by

$$\bar{\rho}_n = \frac{1}{\nu}\int |a_n|^2 D(|a_n|,t)\delta(1 - \sum_n |a_n|^2)\prod_n |a_n|d|a_n|d\phi_n \tag{20.52}$$

21. Time Dependence of the Density Matrix

It is evident from Eqs.(20.33) and (20.41) that D/ν is the probability density in the quantum phase space for the quantum statistical ensemble

$$\frac{D}{\nu} \equiv \text{probability density in the quantum phase space} \qquad (21.1)$$

The statistical density matrix $\bar{\rho}_{nm}(t)$ in Eq.(20.40) is obtained by integration over this probability distribution. It will be time-dependent, since the distribution function $D(\xi, t)$ itself develops in time according to Ehrenfest's theorem in Eq.(20.39). Let us calculate the time derivative of $\bar{\rho}_{nm}(t)$

$$\frac{d}{dt}\bar{\rho}_{nm}(t) = \frac{1}{\nu} \int \rho_{nm} \left[\frac{\partial D}{\partial t}\right] \delta(1 - \sum_n \xi_n^2) \prod_n d\xi_n \qquad (21.2)$$

Substitution of Eq.(20.39) leads to the expression

$$\frac{d}{dt}\bar{\rho}_{nm}(t) = \frac{1}{\nu} \int \rho_{nm} \left[-\sum_p \frac{\partial D}{\partial \xi_p}\dot{\xi}_p\right] \delta(1 - \sum_n \xi_n^2) \prod_n d\xi_n \qquad (21.3)$$

We may now make use of the fact that the normalization is preserved in time as indicated in Eq.(20.31)

$$\frac{d}{dt}\left[\sum_n \xi_n^2(t)\right] = 0 \qquad (21.4)$$

This implies

$$\frac{d}{dt}\left[\delta(1 - \sum_n \xi_n^2(t))\right] = 0 \qquad (21.5)$$

which states that the analytic expression restricting the integration to the unit sphere is independent of time. Since the delta-function depends on the variables $\{\xi_n(t);\ n = 1, 2, \ldots, \infty\}$, the vanishing of this total time derivative can be restated formally as

$$\sum_p \dot{\xi}_p \frac{\partial}{\partial \xi_p}\left[\delta(1 - \sum_n \xi_n^2(t))\right] = 0 \qquad (21.6)$$

and Eq.(21.3) can thus be rewritten as

$$\frac{d}{dt}\bar{\rho}_{nm}(t) = \frac{1}{\nu} \int \rho_{nm} \left\{-\sum_p \dot{\xi}_p \frac{\partial}{\partial \xi_p}\left[D\,\delta(1 - \sum_n \xi_n^2)\right]\right\} \prod_n d\xi_n \qquad (21.7)$$

This result reflects the freedom of putting the analytic restriction to the unit sphere *either* in the volume element *or* in the distribution function D itself.

The integration in Eq.(21.7) goes over all values of ξ_n, and each of the derivatives with respect to these quantities in Eq.(21.7) can now be *partially integrated*. There will be no surface contributions from $\pm\infty$, since the integrand vanishes everywhere except on the unit sphere. Thus one can write

$$\frac{d}{dt}\bar{\rho}_{nm}(t) = \frac{1}{\nu}\int\left[\sum_p\frac{\partial}{\partial\xi_p}(\dot{\xi}_p\rho_{nm})\right]D(\xi,t)\delta(1-\sum_n\xi_n^2)\prod_n d\xi_n \quad (21.8)$$

This may further be rewritten as

$$\frac{d}{dt}\bar{\rho}_{nm}(t) = \frac{1}{\nu}\int\left[\sum_p\dot{\xi}_p\frac{\partial}{\partial\xi_p}(\rho_{nm})\right]D(\xi,t)\delta(1-\sum_n\xi_n^2)\prod_n d\xi_n \quad (21.9)$$

That is, the $\dot{\xi}_p$ can be taken through the derivative $\partial/\partial\xi_p$. The justification for this is obtained by writing out the Schrödinger Eq.(18.23), which governs the time development of the coefficients, in terms of real and imaginary parts

$$\dot{\alpha}_p = \frac{1}{\hbar}\sum_s[(ImH_{ps})\alpha_s + (ReH_{ps})\beta_s]$$
$$\dot{\beta}_p = \frac{1}{\hbar}\sum_s[(ImH_{ps})\beta_s - (ReH_{ps})\alpha_s] \quad (21.10)$$

Differentiation of these relations then gives

$$\frac{\partial\dot{\alpha}_p}{\partial\alpha_p} = \frac{\partial\dot{\beta}_p}{\partial\beta_p} = \frac{1}{\hbar}(ImH_{pp}) = 0 \quad (21.11)$$

where the last equality holds because the diagonal elements of a hermitian matrix are real. Hence, upon summation over p

$$\sum_p\frac{\partial}{\partial\xi_p}(\dot{\xi}_p\rho_{nm}) = \sum_p\left[\frac{\partial}{\partial\alpha_p}(\dot{\alpha}_p\rho_{nm}) + \frac{\partial}{\partial\beta_p}(\dot{\beta}_p\rho_{nm})\right]$$
$$= \sum_p\left[\dot{\alpha}_p\frac{\partial}{\partial\alpha_p}(\rho_{nm}) + \dot{\beta}_p\frac{\partial}{\partial\beta_p}(\rho_{nm})\right] = \sum_p\dot{\xi}_p\frac{\partial}{\partial\xi_p}(\rho_{nm}) \quad (21.12)$$

and the result in Eq.(21.9) is established.

The expression appearing in Eq.(21.9) can be further analyzed by again writing it out in detail in terms of real and imaginary parts

$$\sum_p \dot{\xi}_p \frac{\partial}{\partial \xi_p}(\rho_{nm})$$

$$= \left(\dot{\alpha}_n \frac{\partial}{\partial \alpha_n} + \dot{\beta}_n \frac{\partial}{\partial \beta_n} + \dot{\alpha}_m \frac{\partial}{\partial \alpha_m} + \dot{\beta}_m \frac{\partial}{\partial \beta_m} \right) (\alpha_m - i\beta_m)(\alpha_n + i\beta_n)$$

$$= (\dot{\alpha}_m - i\dot{\beta}_m)(\alpha_n + i\beta_n) + (\dot{\alpha}_n + i\dot{\beta}_n)(\alpha_m - i\beta_m)$$

$$\tag{21.13}$$

This may be rewritten as

$$\sum_p \dot{\xi}_p \frac{\partial}{\partial \xi_p}(\rho_{nm}) = \dot{a}_m^* a_n + a_m^* \dot{a}_n \quad ; \text{ all } m \text{ and } n \tag{21.14}$$

The reader will readily verify that this result holds for all m and n. The insertion of the Schrödinger Eqs.(18.23) and (18.28) then leads to

$$\sum_p \dot{\xi}_p \frac{\partial}{\partial \xi_p}(\rho_{nm}) = \frac{1}{i\hbar} \sum_p [a_m^* H_{np} a_p - a_p^* H_{pm} a_n]$$

$$= \frac{1}{i\hbar} \sum_p [H_{np}\rho_{pm} - \rho_{np}H_{pm}] \tag{21.15}$$

$$= \frac{1}{i\hbar} [\underline{H}\underline{\rho} - \underline{\rho}\underline{H}]_{nm}$$

Hence we establish the relation

$$-\frac{\hbar}{i} \frac{d}{dt} \bar{\rho}_{nm}(t) = \frac{1}{\nu} \int [\underline{H}, \underline{\rho}]_{nm} D(\xi, t) \delta(1 - \sum_n \xi_n^2) \prod_n d\xi_n \tag{21.16}$$

Now the hamiltonian matrix appearing in this equation is a known expression computed from Eq.(18.24), and the only quantities being averaged over the quantum phase space are the density matrix elements themselves. Hence the hamiltonian matrix comes out of the averaging and we have the result that

$$-\frac{\hbar}{i} \frac{d}{dt} \bar{\rho}_{nm}(t) = [\underline{H}, \underline{\bar{\rho}}(t)]_{nm} \tag{21.17}$$

This important relation states that the time development of the statistical density matrix is given by the commutator of that matrix with the hamiltonian matrix. It allows one to determine the statistical density matrix $\bar{\rho}_{nm}(t)$, given its value $\bar{\rho}_{nm}(t_0)$ at some initial time t_0.

22. Thermal Equilibrium

The expectation value of an observable Q in the quantum statistical ensemble is obtained by taking the trace of the matrix product with the statistical density matrix according to Eq.(20.42)

$$\overline{<Q>} = \text{Trace } (\underline{Q}\bar{\rho}) \tag{22.1}$$

We expect that this quantity will be a constant independent of time for a system in thermal equilibrium, and hence we *define* thermal equilibrium by the expression

$$\frac{d}{dt}\overline{<Q>} = 0 \tag{22.2}$$

Throughout this discussion it will be assumed that the operator Q is arbitrary and has no explicit dependence on the time as in Eqs.(18.25-26). An argument directly analogous to that given in Section 7 then implies that

$$\frac{d}{dt}\bar{\rho}_{nm}(t) = 0 \tag{22.3}$$

Thus the statistical density matrix must be independent of time for an ensemble in thermal equilibrium. It follows from Eq.(21.17) that the statistical density matrix $\bar{\rho}$ must commute with the hamiltonian matrix $\underline{H}$

$$[\underline{H}, \bar{\rho}]_{nm} = 0 \tag{22.4}$$

In summary, for the quantum statistical ensemble in thermal equilibrium, the statistical density matrix must be a constant of the motion.

One possible solution to Eq.(22.4), which in fact will again turn out to be our basic result, is that the statistical density operator is a function of the hamiltonian operator

$$\bar{\rho} = F(H) \tag{22.5}$$

Matrix elements are computed from this expression according to Eqs.(18.24-25). The completeness relation

$$\sum_n \phi_n(q_1, q_2, \ldots, q_f)\phi_n^*(q_1', q_2', \ldots, q_f') = \delta(q_1 - q_1')\ldots\delta(q_f - q_f') \tag{22.6}$$

permits one to write the matrix element of any product of operators in terms of the matrix product

$$(AB\ldots C)_{mn} = \sum_p \sum_q \cdots \sum_r A_{mp}B_{pq}\ldots C_{rn} \tag{22.7}$$

Equation (22.4) then holds, since any matrix commutes with itself. The result in Eq.(22.5) is similar to the relation in Eq.(9.1) obtained in classical statistical mechanics.

Now suppose that one has a very simple quantum system whose energy levels are discrete and non-degenerate.† The diagonal element of the statistical density matrix is the probability of finding a system in the state n

$$\bar{\rho}_{nn} = \text{probability of finding system in state } n \qquad (22.8)$$

$\bar{\rho}_{nn}$ clearly depends on n, but for a system with no degeneracy, the state n is uniquely characterized by the eigenvalue E_n of the hamiltonian H. Thus $\bar{\rho}_{nn}$ can only depend on E_n and one can write

$$\bar{\rho}_{nn} = F(E_n) \qquad (22.9)$$

Furthermore, in a basis where the hamiltonian $\underline{H}$ is diagonal and the energy uniquely characterizes the state, one has

$$H_{mn} = E_n \delta_{mn} \qquad (22.10)$$

The relation in Eq.(22.4) then becomes (with an interchange of dummy indices)

$$(E_m - E_n)\bar{\rho}_{mn} = 0 \qquad (22.11)$$

which implies

$$\bar{\rho}_{mn} = 0 \quad ; \ m \neq n \qquad (22.12)$$

Thus the statistical density matrix must be diagonal in this case, and a combination of Eqs.(22.9) and (22.12) leads to

$$\bar{\rho}_{mn} = F(E_n)\delta_{mn} \qquad (22.13)$$

which is precisely the result obtained from Eq.(22.5). This non-degenerate system is equivalent to the ergodic system with one degree of freedom discussed in the classical case in Eq.(8.11) and Figure 8.1.

But what if there is degeneracy? This is certainly the more realistic situation for systems with many degrees of freedom. In this case, in addition to the energy eigenvalue E_n which we here denote by the state label n, one must introduce additional quantum numbers which we shall now denote explicitly by the additional state label σ with $\sigma = 1, 2, \ldots, g$,

---

† For example, a one-dimensional harmonic oscillator where $E_n = \hbar\omega(n + 1/2)$; $n = 1, 2, \ldots, \infty$.

where g is the degeneracy of each energy level.† The index σ refers to other observables which commute with H as discussed in Eqs.(20.3-18) and illustrated in Figure (20.1). In the case of degeneracy, the probability p_n of finding the system with energy E_n irrespective of σ is given by

$$p_n \equiv \sum_{\sigma=1}^{g} |a_{m\sigma}|^2$$

$$= \text{ probability of finding the system with energy } E_n \text{ irrespective of } \sigma$$

$$(22.14)$$

Evidently

$$\sum_n p_n = 1 \qquad (22.15)$$

In the basis of energy eigenstates discussed in Eqs.(20.3) to (20.18) and Fig.(20.1), the hamiltonian matrix is diagonal and in the present notation takes the form

$$H_{m\mu,n\sigma} = E_n \delta_{mn} \delta_{\mu\sigma} \quad ; \ \mu, \sigma = 1, 2, \ldots, g \qquad (22.16)$$

Equation (22.4) then implies that

$$(E_m - E_n)\bar{\rho}_{m\mu,n\sigma} = 0 \quad ; \ \mu, \sigma = 1, 2, \ldots, g \qquad (22.17)$$

The statistical density matrix must therefore be *block diagonal*; it will not connect states of different energy. As before, the probability of finding the system in the state (n_σ) is given by the diagonal element of the statistical density matrix

$$\bar{\rho}_{n\sigma,n\sigma} = \text{ probability of finding the system in the state } (n\sigma) \quad (22.18)$$

In contrast to the non-degenerate case, however, these diagonal elements are *not* uniquely characterized by the energy E_n, since the value of σ must also be specified.

Suppose the energy E_n is specified, so that one is restricted to a particular degenerate subspace with dimension $\sigma = 1, 2, \ldots, g$. It follows from Eqs.(18.15-17) that the expansion coefficients in the wave function in Eq.(18.19) must have the form

$$a_{n\sigma}(t) = c_{n\sigma} e^{-iE_n t/\hbar} \qquad (22.19)$$

---

† A more precise notation, which soon proves too cumbersome, is σ_n with $\sigma_n = 1, 2, \ldots, g_n$.

In this case, we *cannot* say that $\bar{\rho}$ is determined only as a function of E_n. We must discuss the behavior of the statistical density matrix in these degenerate subspaces, and the situation is directly analogous to our discussion of quasi-ergodic behavior and sufficiently quasi-ergodic coarse-grain averages in Section 8. A *fine-grain* description of the system would mean that the coefficients $c_{n\sigma}$ are exactly known. For a complex system with a vast number of degrees of freedom, this is again only possible, in principle, with a prohibitive set of measurements.

Let us make the situation more realistic. Any real physical system will always have some small interactions with the external world not included in H. We therefore write

$$H = H_0 + h \tag{22.20}$$

where h is some arbitrary, small perturbing hamiltonian. The hamiltonian matrix is then modified to the form

$$H_{m\mu,n\sigma} = E_n \delta_{\mu\sigma} \delta_{mn} + h_{m\mu,n\sigma} \tag{22.21}$$

There will thus be a small coupling energy between the states of a degenerate subspace, and this implies a consequent *mixing* of the states. Time-dependent perturbation theory indicates how these states are mixed. Within this subspace, the equation of motion for the coefficients $c_{n\sigma}$ defined in Eq.(22.19) follows as in Eq.(18.22)

$$-\frac{\hbar}{i}\frac{d}{dt}c_{n\sigma}(t) = \sum_{\mu=1}^{g} h_{n\sigma,n\mu}c_{n\mu}(t) \quad ; \sigma = 1,2,\ldots,g \tag{22.22}$$

This is a set of coupled first-order differential equations; the value of the coefficients $c_{n\sigma}(t)$ at the time t is determined, given their values $c_{n\sigma}(t_0)$ at an initial time t_0. Now the coefficients $c_{n\sigma}$ with $\sigma = 1,2,\ldots,g$ form a set of g complex numbers, and therefore we can identify the $c_{n\sigma}$ with a real vector (constructed from the real and imaginary parts) in a $2g$-dimensional space. Thus we define[†]

$$c_{n\sigma} \equiv \alpha_{n\sigma} + i\beta_{n\sigma} \quad ; \sigma = 1,2,\ldots,g \tag{22.23}$$

The normalization condition in Eq.(22.14) becomes

$$\sum_{\sigma=1}^{g}(\alpha_{n\sigma}^2 + \beta_{n\sigma}^2) = p_n \tag{22.24}$$

---

† We use the same notation as in Eq.(20.44), although, $a_{n\sigma} = c_{n\sigma}\exp\left(-iE_n t/\hbar\right)$ here differ by a common phase.

This $2g$-dimensional vector is thus confined to lie on a sphere of radius $\sqrt{p_n}$.

As time goes on, the vector $(\alpha_{n\sigma}, \beta_{n\sigma}; \sigma = 1, 2, \ldots, g)$ describes a phase orbit on the sphere; the Schrödinger equation describes the rotation of this vector as indicated in Figure 22.1.

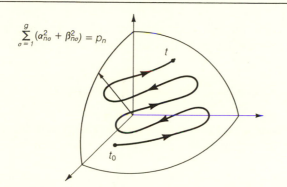

$$\sum_{\sigma=1}^{g} (\alpha_{n\sigma}^2 + \beta_{n\sigma}^2) = p_n$$

Figure 22.1 Schematic illustration of the phase orbit of the vector $(\alpha_{n\sigma}, \beta_{n\sigma}; \sigma = 1, 2, \ldots, g)$ defined in Eqs.(20.23) on the sphere of radius $\sqrt{p_n}$ in the degenerate subspace with energy E_n. The time development in the degenerate subspace is governed by Eq.(22.22).

It was established in Eq.(20.38) that the distribution function D is a constant along a phase orbit. If one could show that D was in fact constant over the entire sphere, it could then be concluded that $D = D(p_n)$, and since p_n is determined only by E_n, we could then write equivalently $D = D(E_n)$ and immediately derive the analog of the relation satisfied by the classical distribution function in Eq.(8.2). Equation (22.5) would then follow directly, as we shall demonstrate below. We cannot say that D is the same for all points on the sphere, however, unless the phase orbit passes through every point on the sphere (i.e. the system is *ergodic*). In fact, it is sufficient for a description of macroscopic systems in thermal equilibrium to make a *coarse-grain description* of the system. We will subdivide the sphere into small areas over which the physical properties of macroscopic systems are sensibly the same, and then assume that the phase orbit comes sufficiently close to representative points in each of these areas so that the distribution function is *effectively* a constant over the sphere (i.e. we assume a *sufficiently-ergodic, coarse-grain* description). The argument is the direct analog of that given in Section 8 in the classical case.

We will thus say that $D(\alpha_{n\sigma}, \beta_{n\sigma})$ is a *constant* for all values of $(\alpha_{n\sigma}, \beta_{n\sigma}; \ \sigma = 1, 2, \ldots, g)$ such that the sum of the squares is given by Eq.(22.24). We also know from Eq.(22.17) that since the statistical density matrix is block diagonal

$$\bar{\rho}_{m\mu,n\sigma} \ = \ 0 \quad \text{for } m \neq n \tag{22.25}$$

we may concentrate on the evaluation of $\bar{\rho}_{n\mu,n\sigma}$ in the degenerate subspace. The relevant integral over the distribution function is given by

$$\bar{\rho}_{n\mu,n\sigma} \ \propto \ \int a_{n\sigma}^{*} a_{n\mu} D(\alpha_{n\sigma}, \beta_{n\sigma}) \prod_{\sigma=1}^{g} d\alpha_{n\sigma} d\beta_{n\sigma} \tag{22.26}$$

where the integration region is restricted by Eq.(22.24) to lie on the sphere indicated in Fig. 22.1. Substitution of Eqs.(22.19) leads to

$$\bar{\rho}_{n\mu,n\sigma} \ \propto \ \int c_{n\sigma}^{*} c_{n\mu} D(\alpha_{n\sigma}, \beta_{n\sigma}) \prod_{\sigma=1}^{g} d\alpha_{n\sigma} d\beta_{n\sigma} \tag{22.27}$$

The transformation to polar coordinates in Eqs.(20.44-49) then gives

$$\bar{\rho}_{n\mu,n\sigma} \ \propto \ \int |c_{n\sigma}||c_{n\mu}|e^{i(\phi_{n\mu} - \phi_{n\sigma})} D(|c_{n\sigma}|, \phi_{n\sigma}) \prod_{\sigma=1}^{g} |c_{n\sigma}| d|c_{n\sigma}| d\phi_{n\sigma} \tag{22.28}$$

In general, the distribution function will depend on both the moduli and phases of the expansion coefficients

$$D = D(|c_{n\sigma}|, \phi_{n\sigma}) \tag{22.29}$$

But now, under the assumption that the sytem is sufficiently quasi-ergodic, one can choose a single, constant distribution on each energy surface. The distribution function therefore depends only on the combination

$$D = D(\sum_{\sigma=1}^{g} |c_{n\sigma}|^2) = D(p_n) \tag{22.30}$$

It is *independent of the phases*, and hence just as in Eqs.(20.50-52), one concludes that

$$\bar{\rho}_{n\mu,n\sigma} = \bar{\rho}_{n\sigma,n\sigma} \delta_{\mu\sigma} \quad ; \ \mu, \sigma = 1, 2, \ldots, g \tag{22.31}$$

But now we can go further. The remaining integral with Eq.(22.30) is the same for all $|c_{n\sigma}|$, because integration over the sphere defined by

$$\sum_{\sigma=1}^{g} |c_{n\sigma}|^2 = p_n \qquad (22.32)$$

does not distinguish them. Thus we arrive at

$$\bar{\rho}_{n\mu,n\sigma} = \bar{\rho}_{nn}\delta_{\mu\sigma} \qquad (22.33)$$

Since $\bar{\rho}_{nn}$ is now only a function of n, and since there is a one-to-one correspondence between n, p_n, and E_n, we can again take the statistical density matrix to be precisely specified by the energy E_n.

Thus, in summary, in a non-degenerate (ergodic) system, it is clear that the statistical density matrix $\underline{\bar{\rho}}$ is diagonal and depends only on the energy

$$\bar{\rho}_{mn} = F(E_n)\delta_{mn} \qquad (22.34)$$

In a degenerate (quasi-ergodic) system, the statistical density matrix is again diagonal, but in addition, it takes on *equal values* for an entire degenerate subspace

$$\bar{\rho}_{m\mu,n\sigma} = F(E_n)\delta_{mn}\delta_{\mu\sigma} \quad ; \ \mu,\sigma = 1, 2, \ldots, g \qquad (22.35)$$

We have thus arrived at the central result, that for the quantum statistical ensemble in thermal equilibrium, provided that the system is sufficiently quasi-ergodic, we can write the operator relation

$$\bar{\rho} = F(H) \qquad (22.36)$$

The statistical density operator must be a function only of the hamiltonian as in Eq.(22.5). The matrix elements of this operator in the basis of energy eigenstates discussed in Eqs.(20.3-20.18) then lead precisely to Eq.(22.35).

It remains to determine the *form* of the function F in Eq.(22.36). This will be done in the next section, as before, by considering a system in thermal equilibrium with other systems.

Before leaving this section, we simply give the analogues of Eq.(8.14) and Figure 8.4 for the quantum statistical ensemble. Equation (22.17) states quite generally that in the energy representation the statistical density matrix is block diagonal and will not connect states of different energy. For the thermodynamic functions of experimental interest, the assumptions of "sufficiently quasi-ergodic behavior" and "coarse-grain averaging" permit the multiple integral in Eq.(20.48) for the statistical density matrix to be written as

$$\bar{\rho}_{n\mu,n\sigma} = \frac{1}{\nu} \sum_{\{P\}} a_{n\sigma}^* a_{n\mu} D(\alpha_{n\sigma}, \beta_{n\sigma}) \delta(1 - \sum_n p_n^2) \prod_n \prod_{\sigma=1}^g \Delta\alpha_{n\sigma} \Delta\beta_{n\sigma}$$

$$(22.37)$$

Here the sum over the set of points $\{P\}$ goes first over points which lie on a finite segment of an arbitrarily chosen phase orbit on a given energy surface with specified E_n, n, and p_n and then over the distinct energy surfaces (Figure 22.2). The validity of this replacement can, again, be checked in specific examples. Equations (22.27-36) evidently hold if this replacement is justified, and we shall henceforth assume this to be the case.

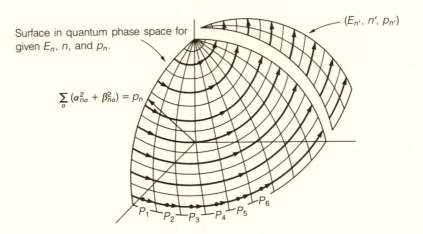

Surface in quantum phase space for given E_n, n, and p_n.

$(E_{n'}, n', p_{n'})$

$\sum_\sigma (\alpha_{n\sigma}^2 + \beta_{n\sigma}^2) = p_n$

P_1 P_2 P_3 P_4 P_5 P_6

Figure 22.2 Schematic illustration of "sufficiently ergodic coarse-grain" averaging in the quantum phase space. The surface corresponds to a given energy E_n, state label n, and probability p_n. The phase orbit is governed by Eq.(22.22).

23. The Canonical Distribution

It was argued in the preceding section that the statistical density operator can be expressed as

$$\bar{\rho} = F(H) \tag{23.1}$$

In order to determine the form of the function F, consider two distinct systems A and B prepared by bringing them into thermal equilibrium with a common heat bath. The argument now proceeds directly as that given in the classical case in Sec. 9. In the quantum case the system A will be characterized by a set of quantum numbers $(n_a \sigma_a)$ and B by $(n_b \sigma_b)$, and the energies will be denoted by $E_{n_a}^{(A)}$ and $E_{n_b}^{(B)}$, respectively, where the notation of the preceding section is employed. The diagonal elements of the statistical density matrix give the probability of finding a system in a particular state. The probability of finding system A in one of the states in the degenerate subspace with energy $E_{n_a}^{(A)}$, and the similar expression for system B, are therefore given by

$$\bar{\rho}_{n_a \sigma_a, n_a \sigma_a}^{(A)} = F_A(E_{n_a}^{(A)})$$
$$= \text{ probability of finding system } A \text{ in state } (n_a \sigma_a) \tag{23.2}$$
$$\bar{\rho}_{n_b \sigma_b, n_b \sigma_b}^{(B)} = F_B(E_{n_b}^{(B)})$$
$$= \text{ probability of finding system } B \text{ in state } (n_b \sigma_b) \tag{23.3}$$

The joint probability of finding system A in $n_a \sigma_a$ *and* system B in $n_b \sigma_b$ is the product of probabilities

$$F_A(E_{n_a}^{(A)})F_B(E_{n_b}^{(B)}) = \text{probability of finding system } A \text{ in state } (n_a \sigma_a)$$
$$\textit{and} \text{ system } B \text{ in state } (n_b \sigma_b)$$
$$\tag{23.4}$$

Now imagine that there is some slight coupling between the two systems, as indicated schematically in Figure 23.1. The energies of this combined system may be written as

$$E_{n_c}^{(C)} = E_{n_a}^{(A)} + E_{n_b}^{(B)} + \text{ negligible term} \tag{23.5}$$

The probability of finding the combined system C in one of the states in the degenerate subspace with this energy is evidently

$$\rho_{n_c \sigma_c, n_c \sigma_c}^{(C)} = F_C(E_{n_c}^{(C)})$$
$$= \text{ probability of finding system } C \text{ in state } (n_c \sigma_c) \tag{23.6}$$

If the two systems A and B are originally placed in thermal equilibrium with a common heat bath, then they will be in thermal equilibrium with one another, and their properties are *unchanged* when they are connected.

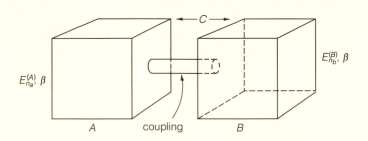

Figure 23.1 Two distinct systems A and B previously placed in thermal equilibrium with a common heat bath. A slight coupling is introduced between them to produce the combined system C, which remains in thermal equilibrium.

This must similarly be true for the ensembles and probability distributions through which we describe them. Thus we demand that the probability of finding a particular state of the combined system C must be the same whether or not the two systems A and B are connected. This implies

$$F_C(E_{n_a}^{(A)} + E_{n_b}^{(B)}) = F_A(E_{n_a}^{(A)})F_B(E_{n_b}^{(B)}) \tag{23.7}$$

The argument now proceeds exactly as in Eqs.(9.13) to (9.22); one concludes that the distribution F must have an exponential form

$$F(E) = Ce^{-\beta E} \tag{23.8}$$

where C is a constant. Here β is again defined to be that quantity which systems in mutual thermal equilibrium have in common [Eq.(9.20)]. The statistical density matrix in the energy representation in Eq.(22.35) thus takes the form

$$\bar{\rho}_{m\mu,n\sigma} = Ce^{-\beta E_n}\delta_{mn}\delta_{\mu\sigma} \tag{23.9}$$

The normalization condition in Eq.(19.4) determines the constant C

$$C^{-1} = \sum_{n,\sigma} e^{-\beta E_n} \tag{23.10}$$

Thus the statistical density matrix in the energy representation is finally given by the *canonical distribution*

$$\bar{\rho}_{m\mu,n\sigma} = \frac{e^{-\beta E_n}}{\sum_{n,\sigma} e^{-\beta E_n}}\delta_{mn}\delta_{\mu\sigma} \tag{23.11}$$

It satisfies

$$\text{Trace } (\bar{\rho}) = \sum_{n,\sigma} \bar{\rho}_{n\sigma,n\sigma} = 1 \qquad (23.12)$$

The argument here has been presented in the energy representation where the hamiltonian matrix H is diagonal. The discussion in Eqs.(22.5-7) allows us to write the final result in an operator form which is independent of any particular representation. The statistical density operator $\bar{\rho}$ corresponding to Eq.(23.11) is evidently

$$\bar{\rho} = \frac{e^{-\beta H}}{\text{Trace } (e^{-\beta H})} \qquad (23.13)$$

It satisfies

$$\text{Trace } (\bar{\rho}) = 1 \qquad (23.14)$$

The thermodynamic average of any dynamical quantity in the *canonical ensemble* follows from Eqs.(19.7) and (20.42)

$$\begin{aligned}
\overline{<Q>} &= \text{Trace } (Q\bar{\rho}) \\
&= \frac{\text{Trace } (Qe^{-\beta H})}{\text{Trace } (e^{-\beta H})} = \frac{\sum_{n,\sigma} Q_{n\sigma,n\sigma} e^{-\beta E_n}}{\sum_{n,\sigma} e^{-\beta E_n}}
\end{aligned} \qquad (23.15)$$

The weighting function is again just the familiar Boltzmann factor $\exp{(-\beta E_n)}$.

24. Thermodynamic Functions and the Partition Function

As in Sec. 10, the mean energy of a system in the canonical ensemble is given by

$$E = \overline{<H>} = \frac{\text{Trace } (He^{-\beta H})}{\text{Trace } (e^{-\beta H})} \qquad (24.1)$$

This expression may be rewritten in the energy representation as

$$\begin{aligned}
E &= \frac{\sum_{n,\sigma} E_n e^{-\beta E_n}}{\sum_{n,\sigma} e^{-\beta E_n}} = \frac{-\frac{\partial}{\partial\beta} \sum_{n,\sigma} e^{-\beta E_n}}{\sum_{n,\sigma} e^{-\beta E_n}} \\
&= -\frac{\partial}{\partial\beta} ln \left(\sum_{n,\sigma} e^{-\beta E_n} \right)
\end{aligned} \qquad (24.2)$$

We define the *quantum partition function* by the expression

$$\mathcal{Z} \equiv \sum_{n,\sigma} e^{-\beta E_n} \equiv \text{Trace } (e^{-\beta H}) \qquad (24.3)$$

The mean energy can then be written in terms of the partition function as

$$E = -\frac{\partial}{\partial \beta} ln \mathcal{Z} \tag{24.4}$$

which is precisely Eq.(10.7). Since the hamiltonian still depends on external parameters, the partition function in Eq.(24.3) will have the functional form of Eq.(10.6)

$$\mathcal{Z} = \mathcal{Z}(\beta, a_j) \tag{24.5}$$

and the ensemble average of the work done on the system as these external parameters are changed is again given by Eqs.(10.13-14)

$$\delta W = \sum_j \overline{\left(\frac{\partial H}{\partial a_j}\right)} da_j = -\frac{1}{\beta} \sum_j \frac{\text{Trace}\left(\frac{\partial}{\partial a_j} e^{-\beta H}\right)}{\text{Trace}\left(e^{-\beta H}\right)} da_j \tag{24.6}$$

It follows that

$$\delta W = -\frac{1}{\beta} \sum_j \frac{\partial ln \mathcal{Z}}{\partial a_j} da_j \tag{24.7}$$

which is precisely Eq.(10.16).

The entropy is again defined by combining the first law of thermodynamics

$$\delta Q = dE - \delta W \tag{24.8}$$

with the second law of thermodynamics, which holds for quasistatic, reversible thermodynamic processes,

$$\frac{\delta Q}{T} = dS \qquad ; \text{ reversible processes} \tag{24.9}$$

to give

$$TdS = dE - \delta W \tag{24.10}$$

The argument now proceeds exactly as in Eqs.(10.17) to (10.40). Everything is the same as before except for the new definition of the quantum partition function in Eq.(24.3); thus one arrives at Eq.(10.40)

$$\beta = \frac{1}{k_B T} \tag{24.11}$$

where k_B is Boltzmann's constant of Eq.(10.44). The entropy is given by Eq.(10.47)

$$S = k_B \left\{ ln \mathcal{Z} - \beta \frac{\partial ln \mathcal{Z}}{\partial \beta} \right\} + S_0 \tag{24.12}$$

where S_0 is a constant independent of β and a_j.

The Nernst Heat Theorem. The theorem of Nernst states that as the temperature $T \longrightarrow 0$, the entropy of any system becomes a universal constant; one is free to define this constant to be zero and hence

$$\underset{T \longrightarrow 0}{\text{Lim }} S \ = \ 0 \tag{24.13}$$

The theorem also states that the constant-volume heat capacity of the system must vanish in this same limit

$$\underset{T \longrightarrow 0}{\text{Lim }} C_v \ = \ 0 \tag{24.14}$$

Since the vanishing of the heat capacity in this limit implies that it becomes increasingly difficult to transfer heat from one substance to another at very low temperatures, and that any heat flow to a sample, no matter how small, will raise the temperature of that sample, the Nernst theorem may also be characterized by the statement, "One can never reach absolute zero."

We proceed to demonstrate this result. First observe that the quantity S_0 in Eq.(24.12) is a constant independent of β and a_j. Since only entropy differences are measured in thermodynamics, one is free to define this constant to have any value; it is defined to be zero in quantum statistical mechanics

$$S_0 \equiv 0 \tag{24.15}$$

Consider the remaining expression in Eq.(24.12) in the limit $T \longrightarrow 0$. The quantum partition function is given by Eq.(24.3)

$$\mathcal{Z} = \sum_{n,\sigma} e^{-\beta E_n} \tag{24.16}$$

Any stable system in quantum mechanics has a lowest energy state, or ground-state, with energy E_0

$$E_0 \equiv \text{ground-state energy} \tag{24.17}$$

We assume here that the ground-state is *non-degenerate*. Let Δ_n be the energy difference between the ground state and the excited states with energy E_n

$$\Delta_n \equiv E_n - E_0 \tag{24.18}$$

The partition function can then be written identically as

$$\mathcal{Z} = e^{-\beta E_0} \sum_{n=0}^{\infty} \sum_{\sigma} e^{-\beta \Delta_n} \tag{24.19}$$

As $T \longrightarrow 0$, β becomes very large, in which case only the first few terms in this series will contribute to the sum. Assume that it takes a finite amount of energy, however small, to excite the first excited state (Figure 24.1).

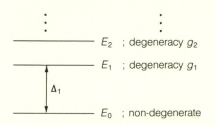

E_2 ; degeneracy g_2

E_1 ; degeneracy g_1

Δ_1

E_0 ; non-degenerate

Figure 24.1 Energy spectrum of quantum system assumed in proof of Nernst heat theorem; the ground-state is assumed to be non-degenerate.

Now take

$$\frac{\Delta_1}{k_B T} \equiv \beta \Delta_1 \gg 1 \tag{24.20}$$

which will be satisfied if T is sufficiently small

$$T \ll \frac{\Delta_1}{k_B} \tag{24.21}$$

In this case the partition function becomes

$$\mathcal{Z} = e^{-\beta E_0} [1 + \sum_{\sigma} e^{-\beta \Delta_1} + \ldots] \tag{24.22}$$

The logarithm of this expression can be written

$$ln\mathcal{Z} = -\beta E_0 + ln[1 + \sum_{\sigma} e^{-\beta \Delta_1} + \ldots]$$

$$\cong -\beta E_0 + \sum_{\sigma} e^{-\beta \Delta_1} \tag{24.23}$$

This approximation improves as $T \longrightarrow 0$. In this limit, the entropy in Eq.(24.12) takes the form

$$S \cong k_B(-\beta E_0 + \sum_\sigma e^{-\beta \Delta_1} + \beta E_0 + \beta \Delta_1 \sum_\sigma e^{-\beta \Delta_1})$$
$$= k_B(1 + \beta \Delta_1) \sum_\sigma e^{-\beta \Delta_1}$$

(24.24)

or

$$S \cong \left(k_B + \frac{\Delta_1}{T}\right) \sum_\sigma e^{-\Delta_1/k_B T}$$

(24.25)

Clearly as $T \longrightarrow 0$, the exponential vanishes sufficiently rapidly so that one is left with

$$\underset{T \longrightarrow 0}{\text{Lim}} \ S \ = \ 0$$

(24.26)

which is the first part of the Nernst theorem.

For temperatures low enough that Eq.(24.21) is satisfied, the entropy will be given by Eq.(24.25). The constant volume heat capacity is defined by

$$C_V \equiv \left(\frac{\partial Q}{\partial T}\right)_V = T \left(\frac{\partial S}{\partial T}\right)_V$$

(24.27)

At constant volume Δ_n does not change. Differentiation of Eq.(24.25) then yields

$$\left(\frac{\partial S}{\partial T}\right)_V \cong \left(-\frac{\Delta_1}{T^2}\right) \sum_\sigma e^{-\Delta_1/k_B T} + \left(k_B + \frac{\Delta_1}{T}\right) \left(\frac{\Delta_1}{k_B T^2}\right) \sum_\sigma e^{-\Delta_1/k_B T}$$

(24.28)

so that

$$T \left(\frac{\partial S}{\partial T}\right)_V \cong k_B \left[\left(\frac{\Delta_1}{k_B T}\right)^2 \sum_\sigma e^{-\Delta_1/k_B T}\right]$$

(24.29)

The exponential again determines the low-temperature behavior, and the heat capacity must indeed vanish in the limit as $T \longrightarrow 0$

$$\underset{T \longrightarrow 0}{\text{Lim}} \ C_V \ = \ 0$$

(24.30)

which is the second part of the Nernst theorem.

Note that these are *quantum* effects because they depend crucially on the availability of a *lowest* energy state and the fact that the energy levels of a quantum system are discrete. Thus, at low enough temperature, all systems will cluster into a single state.   Classically, there is no discreteness, $\Delta_1 = 0$, and the Nernst theorem does not hold.   Indeed, the constant-volume heat capacity of a three-dimensional solid in classical statistical mechanics, for example, satisfies the law of Dulong and Petit in Eq.(16.172c); it is finite and independent of temperature at all temperatures.

VII APPLICATIONS OF QUANTUM STATISTICS

25. Ideal Monatomic Gases

The transition from classical to quantum statistical mechanics is accomplished by the replacement of the classical partition function of Eq.(11.1)

$$\mathcal{Z} = \int e^{-H/k_BT} d\lambda \qquad (25.1)$$

by the quantum partition function of Eq.(24.3)

$$\mathcal{Z} = \text{Trace} \left(e^{-H/k_BT} \right) \qquad (25.2)$$

In the energy representation, the hamiltonian is diagonal

$$H_{mn} = E_n \delta_{mn} \qquad (25.3)$$

and the quantum partition function takes the form

$$\mathcal{Z} = \sum_{\{n\}} e^{-E_n/k_BT} \qquad (25.4)$$

Here we initially revert to the notation of Sec. 20. If the macroscopic system is composed of N *independent subsystems*, then the hamiltonian is a sum of individual contributions as in Eq.(13.2)

$$H = \sum_{s=1}^{N} h_s \qquad (25.5)$$

The energy is similarly a sum of individual contributions

$$E = \sum_{s=1}^{N} \varepsilon_s \qquad (25.6)$$

A state of the macroscopic system is now determined by specifying the set of quantum numbers $\{n\} = \{n_1, n_2, \ldots, n_N\}$, where n_i refers to the ith subsystem. The partition function then takes the form

$$\mathcal{Z} = \sum_{\{n_1\}} \cdots \sum_{\{n_N\}} e^{-(\varepsilon_{n_1} + \ldots + \varepsilon_{n_N})/k_BT} \qquad (25.7)$$

It evidently *factors*

$$\mathcal{Z} = \left(\sum_{\{n_1\}} e^{-\varepsilon_{n_1}/k_B T} \right) \left(\sum_{\{n_2\}} e^{-\varepsilon_{n_2}/k_B T} \right) \cdots \left(\sum_{\{n_N\}} e^{-\varepsilon_{n_N}/k_B T} \right)$$

$$(25.8a)$$

Furthermore, if the subsystems are *identical*, the partition function takes a form identical to that in Eq.(13.7)

$$\mathcal{Z} = z^N \qquad (25.8b)$$

where the one-particle quantum partition function is defined by

$$z \equiv \sum_{\{n\}} e^{-\varepsilon_n/k_B T} \qquad (25.9)$$

This expression is the quantum analog of Eq.(13.8).

As our first application of quantum statistical mechanics, consider an ideal monatomic gas composed of identical subsystems with mass m. The hamiltonian for a single subsystem is given by

$$h = \frac{1}{2m}\mathbf{p}^2 = \frac{1}{2m}\left(p_x^2 + p_y^2 + p_z^2\right)$$

$$= -\frac{\hbar^2}{2m}\left(\frac{\partial^2}{\partial x^2} + \frac{\partial^2}{\partial y^2} + \frac{\partial^2}{\partial z^2}\right) \qquad (25.10)$$

We choose to consider a large cubical box of volume $V = L^3$ (Figure 25.1) and apply periodic boundary conditions as in Figure 16.1d.

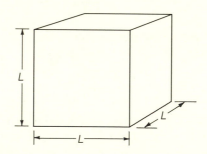

Figure 25.1 Ideal monatomic gas contained in a large box of volume $V = L^3$. Periodic boundary conditions are again assumed as in Figure 16.1d.

In this case, the wavefunctions are just plane waves

$$\phi_{n_x,n_y,n_z}(x,y,z) = \frac{1}{\sqrt{L^3}} e^{i(p_x x + p_y y + p_z z)/\hbar} \tag{25.11}$$

Here (p_x, p_y, p_z) now denote the momentum *eigenvalues*; they are given by

$$(p_x, p_y, p_z) = \frac{2\pi\hbar}{L}(n_x, n_y, n_z) \quad ; \quad n_i = 0, \pm 1, \pm 2, \ldots \pm \infty \tag{25.12}$$
$$i = x, y, z$$

This discrete quantization is imposed by the periodic boundary conditions. The energy eigenvalues are given by

$$\varepsilon_{n_x,n_y,n_x} = \frac{1}{2m}\left(p_x^2 + p_y^2 + p_z^2\right) = \frac{1}{2m}\left(\frac{2\pi\hbar}{L}\right)^2 \left(n_x^2 + n_y^2 + n_z^2\right) \tag{25.13}$$

The one-particle partition function takes the form

$$z = \sum_{n_x}\sum_{n_y}\sum_{n_z} e^{-\beta(n_x^2 + n_y^2 + n_z^2)h^2/2mL^2} \tag{25.14}$$

where, as usual, $\beta \equiv 1/k_B T$. In momentum space, the allowed values of momentum given by Eq.(25.12) form a lattice. This is illustrated in Figure 25.2.

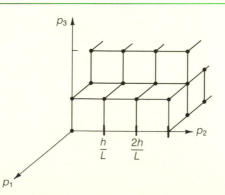

Figure 25.2 Allowed values of the momentum $(p_x, p_y, p_x) = h(n_x, n_y, n_z)/L$ for a particle in a box with periodic boundary conditions.

For large L, the lattice spacing becomes small. Suppose that $d\mathbf{p}$ is a volume in momentum space large compared to the volume of a unit cell. Then, in direct analogy to Eq.(16.168c), the number of quantum states within this volume is given by †

$$\text{\# of quantum states} = \frac{d\mathbf{p}}{(h/L)^3} \tag{25.15}$$

$$= \text{volume in momentum space/volume per lattice site}$$

For a sufficiently large box, the sum in Eq.(25.14) can thus be replaced by an integral

$$z \longrightarrow \frac{L^3}{h^3} \int e^{-\beta \mathbf{p}^2/2m} dp_x dp_y dp_x \tag{25.16}$$
$$V \to \infty$$

The remaining integral is now precisely the classical expression in Eq.(13.11) and the previous analysis applies.‡ The result is that the quantum partition function for a single particle of mass m in a box of volume $V = L^3$ satisfying periodic boundary conditions is given in the limit $V \longrightarrow \infty$ by the expression

$$z \longrightarrow \frac{V}{h^3} \left(\frac{2\pi m}{\beta} \right)^{\frac{3}{2}} \tag{25.17}$$
$$V \to \infty$$

The corresponding translational energy of a system composed of N such subsystems follows as in Eq.(13.22)

$$E = \frac{3}{2} N k_B T \tag{25.18}$$

It agrees with the result obtained from the equipartition theorem. The corresponding molar heat capacity is given by Eq.(13.24)

$$c_v \equiv C_V/\nu_m = \frac{3}{2} R \tag{25.19}$$

One may wonder why this expression does not satisfy the Nernst heat theorem of Eq.(24.30). The answer lies in the fact that the proof

---

† Associate each unit volume with one point, say in the upper righthand corner of the cell.
‡ See, however, Prob. 25.2.

of that result relied on a non-degenerate ground state separated by a finite energy from the first excited state (Figure 24.1). Since we have considered the volume $V = L^3$ to be very large, the first excited state in Eq.(25.13) lies infinitesimally close to the ground state. It is evident from the development leading to Eq.(24.29) that the freezing out of the translational motion then takes place at a temperature only infinitesimally above $T = 0$. Thus, in analogy to Eqs.(16.53b-16.54b)

$$c_v = \frac{3}{2}R \qquad k_BT >> h^2/2mL^2 \qquad (25.20)$$

and the inequality is always satisfied in the limit $V \longrightarrow \infty$.

26. Mean Energy of Harmonic Oscillator

Consider a macroscopic system composed of N identical subsystems which are one-dimensional simple harmonic oscillators. The hamiltonian for a single subsystem is given by

$$h = \frac{1}{2m}p^2 + \frac{1}{2}m\omega^2q^2 = -\frac{\hbar^2}{2m}\frac{d^2}{dq^2} + \frac{1}{2}m\omega^2q^2 \qquad (26.1)$$

It is a basic result of quantum mechanics that the energy spectrum of the one-dimensional simple harmonic oscillator is given by

$$\varepsilon_n = \left(n + \frac{1}{2}\right)\hbar\omega \quad ; \ n = 0, 1, 2, \ldots, \infty$$
$$\equiv \left(n + \frac{1}{2}\right)h\nu \qquad (26.2)$$

This spectrum is shown in Figure 26.1.

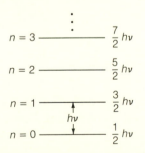

Figure 26.1 Spectrum of the one-dimensional simple harmonic oscillator in quantum mechanics.

The quantum partition function for a single subsystem is defined by

$$z = \sum_n e^{-\beta h\nu(n+ 1/2)} \tag{26.3}$$

where $\beta = 1/k_B T$. The sum appearing in this expression is simply a geometric series

$$z = e^{-\beta h\nu/2} \left[1 + (e^{-\beta h\nu}) + (e^{-\beta h\nu})^2 + \ldots\right] \tag{26.4}$$

which is immediately summed to give

$$z = e^{-\beta h\nu/2} \frac{1}{1 - e^{-\beta h\nu}} \tag{26.5}$$

The partition function for the entire system is again given by Eq.(25.8)

$$\mathcal{Z} = z^N \tag{26.6}$$

and hence

$$ln\mathcal{Z} = Nlnz = N\left[-\frac{\beta h\nu}{2} - ln(1 - e^{-\beta h\nu})\right] \tag{26.7}$$

The mean energy of one oscillator in this system follows from Eq.(24.4)

$$\begin{aligned}\frac{E}{N} &= -\frac{\partial}{\partial\beta}lnz \\ &= \frac{h\nu}{2} + \frac{e^{-\beta h\nu}}{1 - e^{-\beta h\nu}} \cdot h\nu\end{aligned} \tag{26.8}$$

which may be rewritten as

$$\frac{E}{N} = \frac{h\nu}{2} + \frac{h\nu}{e^{\beta h\nu} - 1} \tag{26.9}$$

This is the celebrated *Planck distribution*. It is a *crucial result*. Many of the systems analyzed in the previous applications of classical statistical mechanics were reduced to the problem of a collection of uncoupled simple harmonic oscillators. Equation (26.9) provides the correct expression for the mean energy of a single simple harmonic oscillator of frequency ν in a system composed of many such oscillators, held at temperature T, in quantum statistical mechanics. The first term on the right side of Eq.(26.9) is a constant independent of the temperature. It arises because of the *zero-point energy* in the spectrum in Eq.(26.2). We denote this contribution by E_0 so that

$$E_0 \equiv N(h\nu/2) \qquad (26.10)$$

hence

$$\frac{1}{N}(E - E_0) = \frac{h\nu}{e^{h\nu/k_B T} - 1} \qquad (26.11)$$

Consider the limiting cases of Eq.(26.11). Suppose first that one is interested in *high temperature* so that

$$\beta h\nu \ll 1 \quad \textbf{or} \quad T \gg h\nu/k_B \qquad (26.12)$$

In this case the exponential in the denominator can be expanded in a power series

$$e^{h\nu/k_B T} = 1 + \frac{h\nu}{k_B T} + \dots \qquad (26.13)$$

which yields

$$\frac{1}{N}(E - E_0) = k_B T \qquad (26.14)$$

This is the classical result for the energy of a simple harmonic oscillator; it follows from the equipartition theorem in Eq.(15.13). At high temperature we thus recover the classical result.

Consider next the case of *low temperature* where

$$\beta h\nu \gg 1 \quad \textbf{or} \quad T \ll h\nu/k_B \qquad (26.15)$$

The exponent in Eq.(26.11) now becomes very large, the exponential dominates in the denominator, and one has

$$\frac{1}{N}(E - E_0) = h\nu e^{-h\nu/k_B T} \qquad (26.16)$$

The contribution to the energy in Eq.(26.16) falls exponentially to zero under the conditions of Eq.(26.15). We thus explicitly exhibit the "freezing out" of the excitations at low temperature in quantum statistical mechanics. As discussed in Sec.16, the central failure of classical statistical mechanics was the inability to account for this freezing out of the modes at low temperature as evidenced by the experimental data in a wide variety of physical phenomena.

27. Examples

We proceed to discuss several examples of applications of quantum statistical mechanics. The discussion is based on the analysis of Secs. 16 and 17, where the classical dynamics of several important physical systems was developed: diatomic molecules, solids, radiation in a cavity, and magnetic materials. There, classical statistical mechanics based on the classical hamiltonian and classical partition function was developed, and the successes and inadequacies of that approach emphasized. Here, the systems will be quantized and the quantum partition function evaluated and discussed. Since quantum theory starts from the hamiltonian, however, the hamiltonian dynamics developed in Secs. 16 and 17 provides an immediate point of departure.

a. Diatomic Molecules

Consider a system which is an ideal gas composed of identical diatomic molecules. The hamiltonian for one subsystem follows from Eqs.(16.10b, 16.17b, and 16.37b-16.40b). It has the form

$$h = h_{\text{trans}} + h_{\text{rot}} + h_{\text{vib}} \tag{27.1a}$$

Here

$$h_{\text{trans}} = \frac{1}{2M}\mathbf{P}^2 = \frac{1}{2M}\left(P_x^2 + P_y^2 + P_z^2\right) \tag{27.2a}$$

is the translational kinetic energy and M is the total mass of the molecule [Eq.(16.5b)]. The term h_{rot} is the rotational hamiltonian of Eq.(16.17b)

$$h_{\text{rot}} = \frac{1}{2\mu r_0^2}\left(p_\theta^2 + \frac{p_\phi^2}{\sin^2\theta}\right) \tag{27.3a}$$

As in that equation, it is here assumed that the radial vibrations do not sensibly change the internuclear separation, so that $r \cong r_0$ in this hamiltonian. This is the result for a rigid rotor. The quantity μ is the reduced mass of the molecule [Eq.(16.5b)]. The vibrational hamiltonian is given by

$$h_{\text{vib}} = \frac{1}{2\mu}p_r^2 + \frac{1}{2}\mu\omega^2(r - r_0)^2 \tag{27.4a}$$

where the interatomic potential has been assumed to have the form illustrated in Figure 16.3b, and the quadratic approximation to the potential in the vicinity of the minimum has been retained as in Eqs.(16.37b-16.40b).

Since the hamiltonian of a single subsystem in Eq.(27.1a) separates into a sum of independent contributions under these approximations, the energy of a single subsystem will similarly separate into a sum of independent contributions

$$\varepsilon = \varepsilon_{\text{trans}} + \varepsilon_{\text{rot}} + \varepsilon_{\text{vib}} \tag{27.5a}$$

The partition function for the system now takes the form of Eq.(25.8)

$$\mathcal{Z} = z^N \tag{27.6a}$$

where the partition function for a single subsystem is defined by

$$z = \sum_{\{n\}} e^{-(\varepsilon_{trans}+\varepsilon_{rot}+\varepsilon_{vib})n/k_BT} \tag{27.7a}$$

This expression also *factors*

$$z =$$

$$\left(\sum_{\{n_{\text{trans}}\}} e^{-(\varepsilon_{trans})n/k_BT} \right) \left(\sum_{\{n_{\text{rot}}\}} e^{-(\varepsilon_{rot})n/k_BT} \right) \left(\sum_{\{n_{\text{vib}}\}} e^{-(\varepsilon_{vib})n/k_BT} \right)$$

$$\equiv z_{\text{trans}} \cdot z_{\text{rot}} \cdot z_{\text{vib}} \tag{27.8a}$$

Here $\{n_{\text{trans}}, n_{\text{rot}}, n_{\text{vib}}\}$ stands for the set of quantum numbers appropriate to the various separated problems, which correspond to the different modes of motion of the molecule. The translational partition function z_{trans} for the motion of the center of mass was evaluated in Sec. 25. It is given by

$$z_{\text{trans}} \longrightarrow \frac{V}{h^3} \left(\frac{2\pi M}{\beta} \right)^{3/2} \tag{27.9a}$$
$$V \to \infty$$

We proceed to concentrate on the new contributions arising from z_{rot} and z_{vib}.

The rotational hamiltonian in Eq.(27.3a) can be written as

$$h_{\text{rot}} = \frac{1}{2I} \left(\frac{1}{\sin\theta} p_\theta \sin\theta\, p_\theta + \frac{p_\phi^2}{\sin^2\theta} \right) \tag{27.10a}$$

where I is the moment of inertia of the molecule

$$I \equiv \mu r_0^2 \tag{27.11a}$$

The substitution of Eq.(18.14) now leads to

$$h_{\text{rot}} = \frac{-\hbar^2}{2I} \left(\frac{1}{\sin\theta} \frac{\partial}{\partial\theta} \sin\theta \frac{\partial}{\partial\theta} + \frac{1}{\sin^2\theta} \frac{\partial^2}{\partial\phi^2} \right) \qquad (27.12a)$$

The rewriting of the hamiltonian in Eq.(27.10a) is essential in quantum mechanics so that the resulting differential operator in Eq.(27.12a) is hermitian with respect to the spherical volume element $\sin\theta \, d\theta \, d\phi$ as required by Eqs. (18.24-27).

Now it is a basic result of the quantum theory of angular momentum that if the components of the angular momenta are defined in cartesian coordinates by

$$L_x \equiv (\mathbf{r} \wedge \mathbf{p})_x / \hbar = (1/i) \left(y\frac{\partial}{\partial z} - z\frac{\partial}{\partial y} \right)$$

$$L_y = (1/i) \left(z\frac{\partial}{\partial x} - x\frac{\partial}{\partial z} \right)$$

$$L_z = (1/i) \left(x\frac{\partial}{\partial y} - y\frac{\partial}{\partial x} \right) \qquad (27.13a)$$

with

$$\mathbf{L}^2 \equiv L_x^2 + L_y^2 + L_z^2 \qquad (27.14a)$$

then the following results hold:

1) The components of the angular momentum satisfy the following commutation relations

$$[L_x, L_y] = iL_z \quad ; \text{ and cyclic permutations of } (x, y, z) \qquad (27.15a)$$

2) The square of the total angular momentum commutes with each component of the angular momentum

$$[\mathbf{L}^2, L_i] = 0 \quad ; \; i = x, y, z \qquad (27.16a)$$

in particular

$$[\mathbf{L}^2, L_z] = 0 \qquad (27.17a)$$

so that these operators can be simultaneously diagonalized.

3) The quantities L_z and $\mathbf{L}^2$ can be written in the spherical coordinates of Figure 27.1a as (Prob.27a.3)

$$L_z = (1/i)\frac{\partial}{\partial\phi}$$

$$\mathbf{L}^2 = - \left(\frac{1}{\sin\theta} \frac{\partial}{\partial\theta} \sin\theta \frac{\partial}{\partial\theta} + \frac{1}{\sin^2\theta} \frac{\partial^2}{\partial\phi^2} \right) \qquad (27.18a)$$

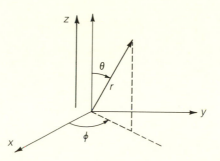

Figure 27.1a Transformation to spherical coordinates.

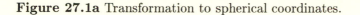

4) The simultaneous eigenfunctions of L_z and $\mathbf{L}^2$ are just the spherical harmonics†, and the corresponding eigenvalues are given by Eqs.(20.15-18)

$$L_z Y_{\ell m_\ell}(\theta, \phi) = m_\ell Y_{\ell m_\ell}(\theta, \phi) \quad ; \quad -\ell \le m_\ell \le \ell$$
$$\mathbf{L}^2 Y_{\ell m_\ell}(\theta, \phi) = \ell(\ell + 1) Y_{\ell m_\ell}(\theta, \phi) \quad ; \quad \ell = 0, 1, 2, \dots, \infty \tag{27.19a}$$

The first few spherical harmonics are given below

$$Y_{00} = \frac{1}{(4\pi)^{1/2}} \; ;$$

$$Y_{11} = -\left(\frac{3}{8\pi}\right)^{1/2} \sin\theta e^{i\phi}$$

$$Y_{10} = \left(\frac{3}{4\pi}\right)^{1/2} \cos\theta$$

$$Y_{1-1} = \left(\frac{3}{8\pi}\right)^{1/2} \sin\theta e^{-i\phi}$$

$$Y_{22} = \frac{1}{4}\left(\frac{15}{2\pi}\right)^{1/2} \sin^2\theta e^{2i\phi}$$

$$Y_{21} = -\left(\frac{15}{8\pi}\right)^{1/2} \sin\theta\cos\theta e^{i\phi}$$

$$Y_{20} = \frac{1}{2}\left(\frac{5}{4\pi}\right)^{1/2}(3\cos^2\theta - 1) \tag{27.20a}$$

$$Y_{2-1} = \left(\frac{15}{8\pi}\right)^{1/2} \sin\theta\cos\theta e^{-i\phi}$$

$$Y_{2-2} = \frac{1}{4}\left(\frac{15}{2\pi}\right)^{1/2} \sin^2\theta e^{-2i\phi}$$

These results can be used to rewrite the hamiltonian in Eq.(27.12a) as

$$h_{\text{rot}} = \frac{\hbar^2}{2I}\mathbf{L}^2 \tag{27.21a}$$

† See L. I. Schiff, *Quantum Mechanics*, 3rd edition, McGraw-Hill, New York (1968).

The eigenfunctions for the rigid rotor are evidently just the spherical harmonics

$$\phi_{\ell,m_\ell}(\theta,\phi) = Y_{\ell m_\ell}(\theta,\phi) \tag{27.22a}$$

and the eigenvalues for the rotational motion of the diatomic molecule follow as

$$\varepsilon_{\rm rot} = \frac{\hbar^2}{2I}\ell(\ell+1) \qquad ; \ell = 0,1,2,\ldots,\infty \tag{27.23a}$$

This characteristic rotational spectrum is sketched in Figure 27.2a.

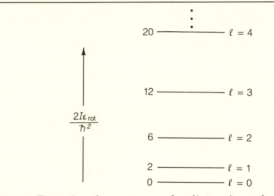

Figure 27.2a Rotational spectrum of a diatomic molecule.

The rotational partition function for the diatomic molecule now follows as

$$z_{\rm rot} = \sum_{\ell,m_\ell} e^{-\beta\hbar^2\ell(\ell+1)/2I} \tag{27.24a}$$

The m_ℓ states are degenerate, since the quantum number m_ℓ does not appear in the rotational eigenvalue. The sum on m_ℓ therefore can be immediately performed to give

$$z_{\rm rot} = \sum_{\ell}(2\ell+1)e^{-\beta\hbar^2\ell(\ell+1)/2I} \tag{27.25a}$$

Consider *high temperature* where

$$\beta\hbar^2/2I \ll 1 \quad \text{or} \quad T \gg \hbar^2/2Ik_B \tag{27.26a}$$

In this high-temperature limit, the levels are closely spaced and the sum in Eq.(27.25a) can be replaced by an integral just as in Sec. 25. If one defines

$$x \equiv \beta \hbar^2 \ell(\ell + 1)/2I$$
$$dx = \beta \hbar^2 (2\ell + 1)/2I \tag{27.27a}$$

then as $\beta \longrightarrow 0$ the rotational partition function takes the form

$$z_{\text{rot}} = \left(\frac{2I}{\beta \hbar^2}\right) \int_0^\infty e^{-x} dx \tag{27.28a}$$

which is immediately evaluated to give

$$z_{\text{rot}} = \frac{1}{\hbar^2} \left(\frac{8\pi^2 \mu r_0^2}{\beta}\right) \tag{27.29a}$$

Apart from the constant factor $1/\hbar^2$, this expression is identical to the classical partition function in Eq.(16.26b). The quantity $T_{\text{rot}}^* \equiv \hbar^2/2Ik_B$ in Eq.(27.26a) which characterizes the high-temperature limit for rotational motion depends on the moment of inertia of the molecule; typically $T_{\text{rot}}^* \sim 100°K$. This high-temperature limit of the present discussion is of particular physical relevance, since most gases are no longer ideal below T_{rot}^*.

At *low temperature*

$$\beta \hbar^2 / 2I >> 1 \quad \text{or} \quad T << \hbar^2/2Ik_B \tag{27.30a}$$

In this case, only the first few terms contribute to the partition function in Eq.(27.25a) so that

$$z_{\text{rot}} \cong 1 + 3e^{-\beta \hbar^2/I} \tag{27.31a}$$

where we retain just the first two terms. The logarithm of this partition function can be approximated by $ln(1 + x) \cong x$ for small x and hence

$$ln z_{\text{rot}} \cong ln(1 + 3e^{-\beta \hbar^2/I}) \cong 3e^{-\beta \hbar^2/I} \tag{27.32a}$$

At these low temperatures, the rotational motion is again "frozen out" just as in the previous discussion of the simple harmonic oscillator in Sec. 26. This discussion of the low-temperature limit really has physical applicability only to hydrogen, since this is the only diatomic molecule that remains a gas at a temperature $T << T_{\text{rot}}^*$.

It is a principle of quantum mechanics that two identical particles with integral spin are *bosons* and that the wave function must be *symmetric* under the interchange of two identical bosons. We say that the wave function must obey *Bose statistics*. Similarly, if two identical particles have half-integral spin they are *fermions* and the wave function must be *antisymmetric* under the interchange of two identical fermions; the wave function obeys *Fermi statistics*.

Homonuclear diatomic molecules made from identical atoms have special restrictions placed on their rotational motion. This arises from the *nuclear statistics* in the following manner. The transformation

$$\theta \longrightarrow \pi - \theta$$
$$\phi \longrightarrow \phi + \pi$$

$(27.33a)$

exhanges the diatomic molecule end-for-end as is evident from Figure 16.1b. It is a property of the spherical harmonics that under this transformation they behave as†

$$Y_{\ell m_\ell}(\pi - \theta, \phi + \pi) = (-1)^\ell Y_{\ell m_\ell}(\theta, \phi)$$

$(27.34a)$

If the two nuclei are identical and have spin zero, as is the case of O_2 with nuclei ^{16}O, then the wavefunction must simply be symmetric under the interchange of the two nuclei and hence ‡

homonuclear diatomic molecule with 0-spin nuclei $\Rightarrow \ell$ even $(27.35a)$

If the nuclei are identical and have spin, then the relevant part of the total wavefunction for the molecule can be written as a product of a rotational wavefunction $Y_{\ell m_\ell}(\theta, \phi)$ and a nuclear spin wavefunction $\chi_{SM_s}(s_1, s_2)$

$$\phi_{\ell m_\ell; SM_s}(\theta, \phi; s_1, s_2) = Y_{\ell m_\ell}(\theta, \phi)\chi_{SM_s}(s_1, s_2)$$

$(27.36a)$

Consider the molecule H_2 whose nuclei are spin-1/2 protons. Two spin-1/2 particles can couple to total spin $S = 1$ (triplet state) or spin $S = 0$ (singlet state). If α denotes the state with spin-up along the z-axis and β the state with spin-down along that axis, then the three triplet states are the familiar combinations

$$\chi_{11} = \alpha(s_1)\alpha(s_2)$$
$$\chi_{10} = \frac{1}{\sqrt{2}}[\alpha(s_1)\beta(s_2) + \beta(s_1)\alpha(s_2)]$$
$$\chi_{1-1} = \beta(s_1)\beta(s_2)$$

$(27.37a)$

---

† This is the same as the parity transformation for a single point particle in a potential well (see Schiff, *op. cit.*).
‡ It is assumed throughout this discussion that the electronic wavefunction is symmetric under Eq.(27.33a).

These three states are evidently *symmetric* in the interchange of the nuclear spins. The wave function for the singlet state is *antisymmetric*

$$\chi_{00} = \frac{1}{\sqrt{2}}[\alpha(s_1)\beta(s_2) - \beta(s_1)\alpha(s_2)] \qquad (27.38a)$$

Since the condition imposed by statistics applies to the behavior of the total wavefunction under the interchange of identical particles, we see that there are two possiblities for H_2

$$S = 0 \text{ (singlet) antisymmetric} \Rightarrow \ell \text{ even } ; \text{ para-hydrogen}$$
$$S = 1 \text{ (triplet) symmetric} \Rightarrow \ell \text{ odd } ; \text{ ortho-hydrogen} \qquad (27.39a)$$

If the nuclei are in the singlet state, one speaks of *para*-hydrogen, and in the triplet state, *ortho*-hydrogen. The rotational partition function is obtained from the sum over all states. For the molecule H_2 it evidently has the form

$$z_{\text{rot}} = \sum_{\ell \text{ even}} (2\ell + 1)e^{-\beta\hbar^2\ell(\ell+1)/2I} \ + 3 \sum_{\ell \text{ odd}} (2\ell + 1)e^{-\beta\hbar^2\ell(\ell+1)/2I}$$

$$\text{(para-hydrogen} \qquad\qquad \text{(ortho-hydrogen}$$
$$\text{contribution)} \qquad\qquad\quad \text{contribution)}$$

$$(27.40a)$$

The factor of 3 in front of the ortho contribution arises from the spin degeneracy of the three states in Eq.(27.37a). For high temperature, the sums reduce to identical integrals and one recovers the classical partition function of Eq.(27.28a) multiplied by an overall, constant, total spin-degeneracy factor of 4 which changes none of our previous thermodynamic results. At low temperature, however, only the first few terms in the sums will contribute, and the partition function depends on the specific nuclear spin states and degeneracies in a non-trivial manner. Here one has

$$z_{\text{rot}} = 1 + 9e^{-2\beta\hbar^2/2I} + 5e^{-6\beta\hbar^2/2I} + \ldots \qquad (27.41a)$$

It follows that the rotational energy of the system is

$$\frac{1}{N}E_{\text{rot}} = -\frac{\partial}{\partial\beta}\ln z_{\text{rot}}$$
$$\cong 18(\hbar^2/2I)e^{-2\beta\hbar^2/2I} \qquad (27.42a)$$

where only the first two terms have been retained in Eq.(27.41a). The rotational contribution to the specific heat of the system is

$$(C_V)_{\text{rot}} = \left(\frac{\partial E_{\text{rot}}}{\partial T}\right)_V = 36Nk_B\left(\frac{\hbar^2}{2Ik_BT}\right)^2 e^{-2\hbar^2/2Ik_BT} \qquad (27.43a)$$

which implies a molar specific heat of

$$(c_v)_{\text{rot}} \equiv \frac{(C_V)_{\text{rot}}}{\nu_m} = 36R \left(\frac{\hbar^2}{2Ik_BT} \right)^2 e^{-2\hbar^2/2Ik_BT} \tag{27.44a}$$

The specific heat of the statistical mixture of ortho- and para-hydrogen discussed above is sketched in Figure 27.3a.

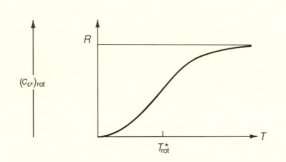

Figure 27.3a Sketch of the molar specific heat of a statistical mixture of ortho- and para-hydrogen. Here $T_{\text{rot}}^* \equiv \hbar^2/2Ik_B$.

The hamiltonian h_{vib} in Eq.(27.4a) describes radial vibrations of the molecule about the equilibrium radius r_0

$$h_{\text{vib}} = -\frac{\hbar^2}{2\mu}\frac{d^2}{dr^2} + \frac{1}{2}\mu\omega^2(r - r_0)^2 \tag{27.45a}$$

The definition

$$q \equiv r - r_0 \tag{27.46a}$$

permits this relation to be rewritten as

$$h_{\text{vib}} = -\frac{\hbar^2}{2\mu}\frac{d^2}{dq^2} + \frac{1}{2}\mu\omega^2 q^2 \tag{27.47a}$$

which is just the hamiltonian of the one-dimensional simple harmonic oscillator in Eq.(26.1). The quantum statistical mechanics of this system has been discussed in Sec. 26. The partition function is given by

$$z_{\text{vib}} = \sum_n e^{-\beta h\nu(n+1/2)} \tag{27.48a}$$

where $h\nu \equiv \hbar\omega$. At high temperature one has from Eq.(26.5)

$$z_{\text{vib}} = \frac{1}{\beta h\nu} = \frac{1}{h}\left(\frac{2\pi k_B T}{\omega}\right) \quad ; \text{ high } T \qquad (27.49a)$$

which is simply $1/h$ times the classical result of Eq.(16.44b). At low temperature one finds from Eq.(26.16)

$$\frac{1}{N}(E - E_0)_{\text{vib}} = h\nu e^{-h\nu/k_B T} \quad ; \text{ low } T \qquad (27.50a)$$

which exhibits the freezing out of the vibrational modes discussed in Sec. 16b and illustrated in Figure 16.4b.

b. Specific Heat of Solids

   The hamiltonian description of the dynamics of a solid was developed in Sec. 16c. For small oscillations about equilibrium, the hamiltonian can always be transformed to normal modes which represent a system of uncoupled simple harmonic oscillators. The determination of the number of normal modes is straightforward. For an isotropic solid of very large volume $V = L^3 \longrightarrow \infty$, the number of modes with wave vector $|\mathbf{k}|$ lying between k and $k + dk$ is given by Eqs.(16.168c) and (16.20d)

$$dn = \frac{4\pi k^2 dk}{(2\pi/L)^3} \quad ; \text{ isotropic solid} \qquad (27.1b)$$

The contribution to the partition function for each normal mode of frequency ν is also readily evaluated, since it follows from the discussion of the simple harmonic oscillator in Sec. 26. The determination of the frequency at which each normal mode with wave number k oscillates, however, is a more complicated problem. It involves the solution of the appropriate dynamical dispersion relation for the system. For example, with the linear chain satisfying periodic boundary conditions it is necessary to solve Eq.(16.135c)

$$\nu(k) = \nu^*\left|\sin\left(\frac{\pi k}{2k^*}\right)\right| \qquad (27.2b)$$

(Here Eq.(16.102c) has been employed for $L \gg a$.) The partition function for a real three-dimensional solid is only obtained by explicitly summing over each of the frequencies solving the appropriate dispersion relation for that system. Fortunately, there exists an approximation to this dispersion relation which provides a remarkably simple and accurate

description of the thermodynamic properties of all solids, in particular of
the specific heat.

The Debye Approximation. Consider the case of low temperature
where all modes but those with the the longest wavelength λ and
correspondingly smallest frequency ν are frozen out. It follows from
the discussion of Eqs.(16.154c) and (16.167c) that in the long-wavelength
limit, the normal modes are just sound waves and the linear dispersion
relation

$$\nu = \frac{k}{2\pi} v_s \qquad (27.3b)$$

may be employed. Here v_s is the sound velocity. This relation implies

$$\frac{k}{2\pi} = \frac{\nu}{v_s} \qquad (27.4b)$$

and hence Eq.(27.1b) may be rewritten directly in terms of frequency as

$$dn = \frac{4\pi V \nu^2 d\nu}{v_s^3} \qquad (27.5b)$$

To be more precise, we should consider not only longitudinal sound waves
with velocity $v_s \equiv v_L$, but also the two transverse modes which propagate
with velocity v_T [we assume these to be the same; see the discussion
following Eq.(16.167c)]. All three of these modes may be incorporated by
defining an average sound velocity v by the relation

$$\frac{3}{v^3} \equiv \frac{1}{v_L^3} + \frac{2}{v_T^3} \qquad (27.6b)$$

Equation (27.5b) may then be rewritten as

$$dn = 3 \cdot \frac{4\pi V \nu^2 dv}{v^3} \qquad (27.7b)$$

The temperature-dependent part of the average energy of an oscillator
of frequency ν is given by the Planck distribution in Eq. (26.11)

$$\bar{\varepsilon}(\nu) = \frac{h\nu}{e^{\beta h\nu} - 1} \qquad (27.8b)$$

The total energy of the solid is obtained by summing this expression
over frequencies. We know, for example from Eq. (16.162c), that there
is a maximum wavenumber k^* and corresponding maximum frequency

ν^* that can propagate through a lattice. Hence this expression will be summed only up to ν^*. The energy of the solid can thus be expressed as

$$E = 3 \cdot \frac{4\pi V}{v^3} \cdot h \int_0^{\nu^*} \frac{\nu^3 d\nu}{e^{\beta h\nu} - 1} \qquad (27.9b)$$

This Debye approximation to the energy holds for low frequency; it thus should provide an accurate low-temperature description of the solid.

Now instead of determining the cut-off frequency from dynamics, let us determine it in another fashion which will turn out to make the Debye approximation a useful representation at *all* temperatures. At high temperature

$$\beta h\nu^* << 1 \quad \text{or} \quad T >> h\nu^*/k_B \qquad (27.10b)$$

Equation (27.9b) should reproduce the classical expression for the energy in this limit; it is given by the equipartition result in Eq.(16.170c)

$$E = 3Nk_BT \qquad (27.11b)$$

In this limit, the integral in Eq.(27.9b) becomes

$$E = 3 \cdot \frac{4\pi V}{v^3} \cdot \frac{1}{\beta} \int_0^{\nu^*} \nu^2 d\nu \qquad (27.12b)$$

Equation of these two expressions yields

$$\frac{4\pi V \nu^{*3}}{3v^3} = N \qquad (27.13b)$$

which allows one to solve for ν^*

$$\nu^* = v \left(\frac{3N}{4\pi V} \right)^{1/3} \qquad (27.14b)$$

The quantity N/V is just the particle density

$$n = N/V \qquad ; \text{ particle density} \qquad (27.15b)$$

and hence Eq.(27.14b) can be rewritten as

$$\nu^* = v \left(\frac{3n}{4\pi} \right)^{1/3} \qquad (27.16b)$$

But $n^{1/3} \sim a^{-1}$ where a is the interatomic distance. Hence this relation states that $\nu^* \sim v/a$, or from Eq. (27.4b)

$$\lambda^* \sim a \qquad (27.17b)$$

This independent determination of λ^* from demanding the proper total number of normal modes at high frequency gives a cut-off wavenumber which is entirely consistent with the dynamical determination of this quantity in Eq.(16.162c)! The Debye approximation to the normal-mode spectrum is compared qualitatively with that of a real solid in Figure 27.1b.

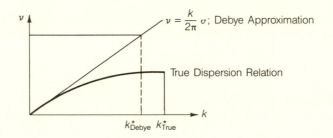

$\nu = \dfrac{k}{2\pi}\,\sigma$; Debye Approximation

True Dispersion Relation

k^*_{Debye} k^*_{True}

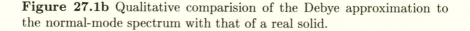

Figure 27.1b Qualitative comparision of the Debye approximation to the normal-mode spectrum with that of a real solid.

At low temperature where

$$\beta h\nu^* >> 1 \quad \text{or} \quad T << h\nu^*/k_B \qquad (27.18b)$$

the upper limit of integration in Eq.(27.9b) can be extended to infinity

$$E = 3 \cdot \frac{4\pi V}{v^3} \cdot h \int_0^\infty \frac{\nu^3 d\nu}{e^{\beta h\nu} - 1} \qquad (27.19b)$$

The introduction of the dimensionless variable

$$x \equiv \beta h\nu = h\nu/k_B T \qquad (27.20b)$$

permits this to be rewritten as

$$E = 3 \cdot \frac{4\pi V}{v^3} \cdot \frac{h}{(\beta h)^4} \int_0^\infty \frac{x^3 dx}{e^x - 1} \qquad (27.21b)$$

The definite integral is given by†

$$\int_0^\infty \frac{x^3 dx}{e^x - 1} = \frac{\pi^4}{15} \qquad (27.22b)$$

At low temperature the Debye approximation thus gives the following expression for the energy

$$E = 3 \cdot \frac{4\pi^5 V}{15 v^3 h^3} (k_B T)^4 = 3 \cdot \frac{\pi^4 N k_B T^4}{5 \, \theta_D^3} \qquad (27.23b)$$

Here we have defined the Debye temperature through the relation

$$\frac{4\pi V}{3 v^3 h^3} \cdot \frac{k_B^3}{N} \equiv \left(\frac{1}{\theta_D}\right)^3 \qquad (27.24b)$$

or

$$\theta_D = \left(\frac{3n}{4\pi}\right)^{1/3} \frac{hv}{k_B} = \frac{h\nu^*}{k_B} \qquad (27.25b)$$

For a typical solid one has

$$v \sim 10^5 \text{ cm/sec} \quad ; \quad \left(\frac{3n}{4\pi}\right)^{1/3} \sim \frac{1}{a} \sim 10^8 /\text{cm} \qquad (27.26b)$$

and hence

$$\theta_D \sim \frac{6 \times 10^{-27} \text{erg sec} \times 10^5 \text{cm/sec}}{1.4 \times 10^{-16} \text{erg}(^\circ K)^{-1}} \times 10^8 /\text{cm} \sim 400^\circ K \qquad (27.27b)$$

Since θ_D is proportional to the velocity of propagation, it is evident from Eqs.(27.6b) and (27.23b) that there will be different Debye temperatures for the longitudinal and transverse modes

$$\frac{3}{\theta_D^3} \equiv \frac{1}{\theta_L^3} + \frac{2}{\theta_T^3} \qquad (27.28b)$$

The energy in Eq.(27.23b) can thus be written in more detail as

$$\frac{E}{N} = \frac{\pi^4 k_B T^4}{5} \left(\frac{1}{\theta_L^3} + \frac{2}{\theta_T^3}\right) \qquad (27.29b)$$

---

† See A. L. Fetter and J. D. Walecka, *Quantum Theory of Many-Particle Systems*, McGraw-Hill Book Co., New York (1971).

The derivative of this expression with respect to temperature at constant volume immediately yields the low-temperature specific heat, which per mole takes the form

$$c_v = \frac{4\pi^4 R}{5}\left[\left(\frac{T}{\theta_L}\right)^3 + 2\left(\frac{T}{\theta_T}\right)^3\right] \qquad ; T << h\nu^*/k_B$$

$$\equiv \frac{12\pi^4 R}{5}\left(\frac{T}{\theta_D}\right)^3$$

(27.30b)

Note that at low temperature the specific heat is proportional to T^3 in this Debye theory. At high temperature, the law of Dulong and Petit in Eq.(16.172c) has now been built into the analysis. The Debye result for the specific heat of a solid is sketched in Figure 27.2b.

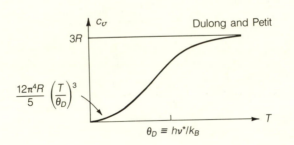

Figure 27.2b Sketch of the specific heat of a solid in the Debye approximation.

The Debye theory of the specific heat agrees well with the specific heat of solids both at high temperature $T >> \theta_L, \theta_T$ and low temperature $T << \theta_L, \theta_T$, although deviations of a few percent depending on crystal structure can be distinguished there. Equations (27.9b) and (27.25b) can be combined to yield the relation

$$\frac{E}{N} = 9(k_B T)\left(\frac{T}{\theta_D}\right)^3 \int_0^{\theta_D/T}\frac{x^3 dx}{e^x - 1}$$

(27.31b)

which provides a remarkably accurate and useful one-parameter representation of the energy and specific heat of a solid at all temperatures.

c. Black-Body Radiation

The hamiltonian dynamics of electromagnetic radiation in a cavity was discussed in Sec. 16d. We can evidently take over the preceding discussion of the quantum theory of the specific heat of solids with the following observations:

1) The dispersion relation for electromagnetic radiation is strictly linear at all frequencies

$$\nu\lambda = c$$

$$\nu = \frac{k}{2\pi}c \qquad (27.1c)$$

The velocity of propagation $v = c$ is now the velocity of light.

2) Owing to the transversality of the electromagnetic fields, there is no longitudinal radiation and only the two transverse modes. The appropriate density of states for black-body radiation then follows from Eqs.(27.5b-27.7b)

$$dn = 2 \cdot \frac{4\pi V \nu^2 d\nu}{c^3} \qquad (27.2c)$$

3) One can have electromagnetic radiation in a cavity with arbitrarily short wavelength, and hence there is no minimum wavelength λ^* or corresponding maximum frequency ν^*. The following limit

$$\nu^* \longrightarrow \infty \qquad (27.3c)$$

of the previous results is therefore appropriate.

As a consequence of observation 3), the relation in Eq.(27.18b) is always satisfied, and one is always in the equivalent *low-temperature* limit for black-body radiation. The energy density in the quantum theory of black-body radiation then follows immediately from Eq.(27.23b)

$$e \equiv \frac{E}{V} = 2\frac{4\pi^5}{15(hc)^3} \cdot (k_B T)^4 \qquad (27.4c)$$

With the introduction of the Stefan-Boltzmann constant

$$a \equiv \frac{8\pi^5 k_B^4}{15(hc)^3}$$

$$= \frac{8\pi^5(1.381 \times 10^{-16} \text{ erg } {}^\circ K^{-1})^4}{15(2\pi \times 1.055 \times 10^{-27} \text{ erg sec} \times 2.998 \times 10^{10} \text{cm/sec})^3} \qquad (27.5c)$$

$$= 7.564 \times 10^{-15} \text{ erg/cm}^3 ({}^\circ K)^4$$

the energy density takes the form

$$e = aT^4 \tag{27.6c}$$

It is also possible to define the energy density per unit frequency from Eq.(27.19b)

$$\frac{de}{d\nu} = \frac{8\pi h}{c^3} \frac{\nu^3}{e^{h\nu/k_B T} - 1} \tag{27.7c}$$

This is the celebrated *Planck distribution*, which we anticipated in Eqs.(16.25d-16.27d) and which is sketched in Figure 27.1c. In the low-frequency limit, one recovers the Rayleigh-Jeans law of Eq.(16.25d)

$$\frac{de}{d\nu} = \frac{8\pi k_B T \nu^2}{c^3} \quad ; \; \nu \ll k_B T/h \tag{27.8c}$$

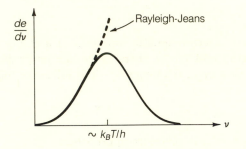

Figure 27.1c Sketch of Planck distribution in Eq.(27.7c) for the energy-density per unit frequency in the quantum theory of black-body radiation.

The Planck distribution provides a quantitative description of the observed black-body spectrum. This agreement provided the original impetus for the development of quantum mechanics.

d. Magnetism

The hamiltonian dynamics of independent subsystems in an external magnetic field was developed in Sec.17. The hamiltonian for an individual subsystem is given by

$$h = h_0 - \boldsymbol{\mu} \cdot \mathbf{H} + \frac{e^2 \mathcal{H}^2}{8mc^2} \rho^2 \tag{27.1d}$$

where

$$\rho^2 \equiv \sum_{s=1}^{Z}(x_s^2 + y_s^2) \qquad (27.2d)$$

The magnetic field can be taken to define the z-axis as indicated in Figure 27.1d. It is then the z-component of the magnetic moment

$$-\boldsymbol{\mu}\cdot\mathbf{H} = -\mu_z\mathcal{H} \qquad (27.3d)$$

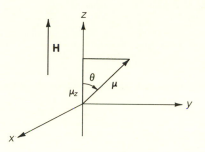

Figure 27.1d Configuration of a quantum mechanical subsystem in an external magnetic field **H**. μ_z is the z-component of the magnetic moment of the subsystem.

which appears in the hamiltonian

$$h = h_0 - \mu_z\mathcal{H} + \frac{e^2\mathcal{H}^2}{8mc^2}\rho^2 \qquad (27.4d)$$

The z-component of the magnetic moment is related to the z-component of the angular momentum $\hbar j_z$ by the expression

$$\mu_z = \gamma\hbar j_z \qquad (27.5d)$$

where γ is the gyromagnetic ratio. Quantum mechanics restricts the z-component of the angular momentum to take the values

$$j_z \equiv m_j \quad ; \quad m_j = -j, -j+1, -j+2, \ldots, +j \qquad (27.6d)$$

and hence

$$\mu_z = \gamma\hbar m_j \qquad (27.7d)$$

The *magnetic moment* μ of the subsystem is defined by giving j_z its maximum value of

$$\mu \equiv \gamma\hbar(j_z)_{\text{max}} = \gamma\hbar j \qquad (27.8d)$$

Equation (27.7d) can thus be rewritten as

$$\mu_z = \mu\frac{m_j}{j} \qquad (27.9d)$$

In the case of *orbital* motion of an electron in an atom, one has $j = \ell$, which takes the integer values $\ell = 0, 1, 2, \ldots$, and the orbital gyromagnetic ratio is given by Eq.(17.38)

$$\gamma_\ell = \frac{-e}{2m_e c} = \frac{-|e|}{2m_e c} \qquad (27.10d)$$

The orbital magnetic moment of the electron is thus given by

$$\mu_\ell = \frac{-e\hbar}{2m_e c}\ell \qquad (27.11d)$$

Its magnitude is characterized by the Bohr magneton of Eq.(17.43)

$$\mu_0 \equiv \frac{e\hbar}{2m_e c} = 9.274 \times 10^{-21} \text{erg/Gauss} \qquad (27.12d)$$

The electron also has internal *spin*. In this case $j = s = 1/2$ and the gyromagnetic ratio for the spin is

$$\gamma_s = -\frac{e}{m_e c} \qquad (27.13d)$$

The spin magnetic moment of the electron is again the Bohr magneton†

$$\mu_s = \frac{-e\hbar}{2m_e c}2s = \frac{-e\hbar}{2m_e c} = -\mu_0 \qquad (27.14d)$$

More generally, the magnetic moment of an electron in an atom has both orbital and spin contributions. The total angular momentum (in units of $\hbar$) is then

$$\mathbf{j} = \boldsymbol{\ell} + \mathbf{s} \qquad (27.15d)$$

---

† The full theory of quantum electrodynamics changes this result slightly.

and the gyromagnetic ratio in Eqs.(27.5d) and (27.8d) now depends on the particular combination of orbital and spin angular momenta

$$\frac{-e}{2m_ec}|\ell + 2s| \leq \frac{\mu}{\hbar} \leq \frac{-e}{2m_ec}|\ell - 2s| \qquad (27.16d)$$

For many-electron atoms, there is an additive contribution from each of the electrons as in Eq.(17.36).

An analogous situation occurs with *nuclei*. There is again a magnetic moment proportional to the angular momentum of the nucleus

$$\mu = \gamma\hbar j \qquad (27.17d)$$

In this case, however, the orbital gyromagnetic ratio involves the nucleon mass, rather than the electron mass

$$\gamma_\ell = \frac{e}{2m_pc} \qquad (27.18d)$$

The nuclear magneton, which characterizes nuclear magnetic effects, is correspondingly reduced

$$\mu_N \equiv \frac{e\hbar}{2m_pc} \cong \frac{1}{1836}\mu_0 \qquad (27.19d)$$

Protons and neutrons both have spin $1/2$, so that $j = s = 1/2$ as with the electron. In contrast, their magnetic moments are substantially different from the nuclear magneton

$$\begin{array}{ll} \text{For protons } \mu_s = 2.793 \ \mu_N & ; \ j = 1/2 \\ \text{neutrons } \mu_s = -1.913 \ \mu_N & ; \ j = 1/2 \end{array} \qquad (27.20d)$$

This indicates that nucleons themselves are composite systems, made up of simpler subsystems. Nuclei composed of many nucleons again receive (basically) additive contributions from the individual subsystems.

Let us denote the ground-state eigenvalue of the hamiltonian h_0 for an individual subsystem by

$$h_0\phi_0 = \varepsilon_0\phi_0 \qquad (27.21d)$$

The ground-state expectation value of the hamiltonian in Eq.(27.4d) is then given by

$$\varepsilon_{m_j} \equiv < \phi_0|h|\phi_0 >$$
$$= \varepsilon_0 - \mathcal{H}\mu\frac{m_j}{j} + \frac{e^2\mathcal{H}^2}{8mc^2} < \rho^2 >_0 \qquad (27.22d)$$

where

$$< \rho^2 >_0 \equiv < \phi_0|\rho^2|\phi_0 > \qquad (27.23d)$$

Equation (27.22d) holds if j is finite, as will be assumed in the following discussion. If $j = 0$, then the second term is absent.

If the thermal and field configurations are such as not to excite the subsystem from its ground state, then the one-particle quantum partition function is obtained by summing over the orientation of the subsystem in the external field

$$z_{\text{int}}(\mathcal{H}) = \sum_{m_j=-j}^{j} e^{-\varepsilon_{m_j}/k_B T} \qquad (27.24d)$$

Hence

$$z_{\text{int}}(\mathcal{H}) = e^{-\beta(\varepsilon_0 + e^2\mathcal{H}^2 <\rho^2>/8mc^2)} \sum_{m_j=-j}^{j} e^{\beta\mathcal{H}\mu m_j/j} \qquad (27.25d)$$

Consider the simplest case, where $j = 1/2$ and m_j takes the two values $+1/2$ and $-1/2$. In this case

$$\sum_{m_j=-j}^{j} e^{\beta\mathcal{H}\mu m_j/j} = e^{\beta\mathcal{H}\mu} + e^{-\beta\mathcal{H}\mu} = 2 \cosh(\beta\mathcal{H}\mu) \qquad (27.26d)$$

The logarithm of the one-particle quantum partition function then takes the form

$$ln\ z_{\text{int}}(\mathcal{H}) = -\beta \left[\varepsilon_0 + \frac{e^2\mathcal{H}^2}{8mc^2} <\rho^2>_0 \right] + ln[\cosh(\beta\mathcal{H}\mu)] + ln2 \qquad (27.27d)$$

For a system of N independent atoms (or nuclei), Eqs.(17.47) and (17.54) imply that

$$lnZ_{\text{int}}(\mathcal{H}) = Nlnz_{\text{int}}(\mathcal{H}) = -\beta F(\mathcal{H}) \qquad (27.28d)$$

or

$$F(\mathcal{H}) = -\frac{1}{\beta}Nlnz_{\text{int}}(\mathcal{H}) \qquad (27.29d)$$

The magnetization follows from Eq.(17.56)

$$M = -\left(\frac{\partial F}{\partial \mathcal{H}}\right)_{V,T} = \frac{N}{\beta}\frac{\partial lnz_{\text{int}}(\mathcal{H})}{\partial \mathcal{H}} \qquad (27.30d)$$

which yields

$$M = -N\frac{e^2\mathcal{H}}{4mc^2} <\rho^2>_0 + N\mu \tanh(\beta\mathcal{H}\mu) \qquad (27.31d)$$

This magnetization can again be separated into a diamagnetic and paramagnetic contribution

$$M \equiv M_{\text{dia}} + M_{\text{para}} \tag{27.32d}$$

The diamagnetic contribution yields the previously discussed diamagnetic susceptibility of Eqs.(17.44) and (17.63). The paramagnetic magnetization is given in this quantum theory by the expression

$$\frac{M_{\text{para}}}{V} = n\mu \tanh(\beta \mathcal{H}\mu) \tag{27.33d}$$

where $n \equiv N/V$ is the particle density as in Eq.(17.59). At low field strengths

$$\beta \mathcal{H}\mu \ll 1 \quad \text{or} \quad \mathcal{H} \ll k_B T/\mu \tag{27.34d}$$

this expression reduces to

$$\frac{M_{\text{para}}}{V} = \frac{n\mu^2}{k_B T}\mathcal{H} \tag{27.35d}$$

which yields the paramagnetic susceptibility

$$\chi_{\text{para}} = \frac{n\mu^2}{k_B T} \tag{27.36d}$$

This result should be compared with the semi-classical result of Eq.(17.62)

$$\chi_{\text{para}} = \frac{1}{3}\frac{n\mu^2}{k_B T} \quad ; \text{ semi-classical} \tag{27.37d}$$

In fact, this latter result is recovered from Eq.(27.25d) in the limit of very large j (Problem 27d.1). At high field strengths

$$\beta \mu \mathcal{H} \gg 1 \quad \text{or} \quad \mathcal{H} \gg k_B T/\mu \tag{27.38d}$$

Eq.(27.33d) implies "saturation"

$$\frac{M_{\text{para}}}{V} = n\,\mu \tag{27.39d}$$

The paramagnetic contribution to the magnetization given by Eq.(27.33d) at all $\mathcal{H}$ is sketched in Figure 27.2d.
This analysis can be extended to provide a simple model of ferromagnetism.

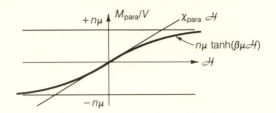

Figure 27.2d Sketch of quantum theory of paramagnetism in Eq.(27.33d).

Weiss Theory of Ferromagnetism. There are two sources of the actual magnetic field inside a long, thin magnetic sample as shown in Figure 17.1. One is the the magnetic field $\mathcal{H}$ arising, for example, from an external solenoid. The second is from the internal magnetization M itself. The presence of both these contributions can be taken into account by replacing

$$\mathcal{H} \longrightarrow \mathcal{H}' \equiv \mathcal{H} + \alpha \frac{M}{V} \qquad (27.40d)$$

where α is a dimensionless constant, whose evaluation involves a calculation of the actual internal field at the location of one of the individual subsystems. The magnetization here is evidently given by

$$\frac{M}{V} = \frac{1}{\alpha}(\mathcal{H}' - \mathcal{H}) \qquad (27.41d)$$

With the incorporation of this modification Eq.(27.33d) now becomes

$$\frac{M}{V} = \frac{1}{\alpha}(\mathcal{H}' - \mathcal{H}) = n\mu \tanh{(\beta \mathcal{H}' \mu)} \qquad (27.42d)$$

This is a transcendental equation for the actual field $\mathcal{H}'$ in terms of the applied field $\mathcal{H}$. It may be solved graphically as indicated in Figure 27.3d.

An essential feature of this result is that there is now the possibility of a *non-vanishing* solution to this equation for *vanishing* applied field $\mathcal{H}$! If this solution occurs in the region where the right side of Eq.(27.42d) has saturated, then

$$\frac{M}{V} = \frac{1}{\alpha}\mathcal{H}' \cong n\mu \quad ; \; \mathcal{H} = 0 \qquad (27.43d)$$

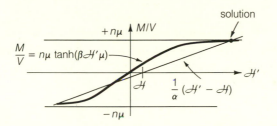

Figure 27.3d Graphical solution of transcendental Eq.(27.41d) for the internal magnetic field in Weiss theory of ferromagnetism.

Typical values here are given by Eq.(17.61) and hence

$$\frac{1}{\alpha}\mathcal{H}' \sim (10^{23}/\text{cm}^3)(10^{-20}\text{erg/Gauss}) = 10^3 \text{ Gauss} \qquad (27.44d)$$

To obtain a typical internal ferromagnetic field of

$$\mathcal{H}' \sim 10^5 \text{ Gauss} \qquad (27.45d)$$

a value of

$$\alpha \sim 10^2 \qquad (27.46d)$$

is required. This Weiss theory provides a very useful phenomenological description of ferromagnetism.

e. Transition from Quantum to Classical Statistical Mechanics

We conclude this section on applications of quantum statistical mechanics with a brief discussion of the transition from quantum to classical statistical mechanics. The quantum partition function in the canonical ensemble is given by Eq.(25.2)

$$\mathcal{Z} = \text{ Trace } (e^{-H/k_B T}) \qquad (27.1e)$$

The classical partition function is given by Eq.(25.1)

$$\mathcal{Z}_{c\ell} = \int e^{-H/k_B T} d\lambda \qquad (27.2e)$$

In the case of N independent subsystems, both expressions take the form

$$\mathcal{Z} = z^N \tag{27.3e}$$

Consider first the example of the ideal monatomic gas. When $k_B T$ is large compared to the level spacing as in Eq.(25.20), then the quantum mechanical partition function can be replaced by an *integral* and the result in Eq.(25.16) gives

$$z \longrightarrow \frac{1}{h^3} z_{c\ell} \tag{27.4e}$$

Consider next an ideal gas of diatomic molecules. Again, when $k_B T$ is large compared to the level spacing as in Eq.(27.26a), the quantum mechanical partition function can be replaced by an integral and Eq.(27.29a) then yields

$$z_{\text{rot}} \longrightarrow \frac{1}{h^2} (z_{\text{rot}})_{c\ell} \tag{27.5e}$$

Similarly, under the high-temperature condition of Eq.(26.12), the vibrational partition function of Eq.(27.49a) becomes

$$z_{\text{vib}} \longrightarrow \frac{1}{h} (z_{\text{vib}})_{c\ell} \tag{27.6e}$$

It is evident from Eq.(27.48a) that this is the general result for a one-dimensional simple-harmonic oscillator.

Since the Debye theory of the specific heat of solids involves $3N$ uncoupled simple harmonic oscillators (the normal modes), it will again be true here that in the high-temperature limit

$$\mathcal{Z} \longrightarrow \frac{1}{h^{3N}} \mathcal{Z}_{c\ell} \tag{27.7e}$$

It is evident from all of these examples that at high temperature the quantum partition function goes over to the classical partition function multiplied by $(1/h)^f$, where f is the number of degrees of freedom for the problem under consideration. We may state this result in the following way

$$\mathcal{Z} \longrightarrow \frac{1}{h^f} \mathcal{Z}_{c\ell}$$

high T

$$= \int e^{-H/k_B T} \prod_{i=1}^{f} \left(\frac{dp_i dq_i}{h} \right) \tag{27.8e}$$

In this form the relation is not only of great utility in practical calculations, it has a very direct interpretation in terms of the discussion of sufficiently ergodic behavior and coarse-grain averaging in Sec. 8. The Heisenberg uncertainty principle in Eq.(18.0) states that p_i and q_i with $i = 1, 2, \ldots, n$ cannot be simultaneously specified more accurately than $\sim h$. One can thus imagine that phase space is divided into *cells of area h* as indicated schematically in Figure 27.1e, and that the only specification of the state of a system which has any physical relevance is whether or not the system is inside of a given cell.† Thus, quantum mechanics implies that a coarse-grain average in phase space is not just a mathematical convenience, but is actually a *physical necessity.*

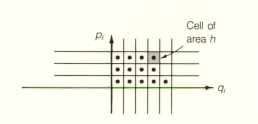

Figure 27.1e Division of phase space into cells of area h and representative points specifying the state of a system.

28. Identical Particles

In the treatment of the quantum theory of ideal gases in Sec. 25, a state of the system was characterized by the momenta $\mathbf{p}_1, \mathbf{p}_2, \ldots, \mathbf{p}_s, \ldots, \mathbf{p}_N$ of particles $1, 2, \ldots s, \ldots, N$. This is permissible if the particles are distinguishable from each other, but new features appear for identical particles where this may no longer be the case. Suppose that one has found a state of a system of identical particles with a given energy E_n. An interchange of particles then results in a state with the *same* energy, for if

$$\phi_n^{(1)}(\mathbf{r}_1, \mathbf{r}_2, \ldots, \mathbf{r}_N) \tag{28.1}$$

† See, in this connection, Probs. 25.2 and 31.4.

is the wavefunction of such a state, so is

$$\phi_n^{(2)}(\mathbf{r}_2, \mathbf{r}_1, \ldots, \mathbf{r}_N) \tag{28.2}$$

and so on. In fact, any linear combination of the $N!$ wavefunctions obtained by permuting the particle labels

$$\phi_n(\mathbf{r}_1, \mathbf{r}_2, \ldots, \mathbf{r}_N) = \sum_{k=1}^{N!} C^{(k)} \phi_n^{(k)}(\mathbf{r}_1, \mathbf{r}_2, \ldots \mathbf{r}_N) \tag{28.3}$$

will again be an eigenstate with the same energy E_n; here $C^{(k)}$ is an arbitrary coefficient. One may thus have a *degeneracy*, possibly of very high order. This degeneracy is actually *removed* in nature. Only two possible linear combinations are actually realized:

Either $C^{(1)} = C^{(2)} = \ldots = C^{(N!)}$ so that

$$\phi_n(\mathbf{r}_1, \mathbf{r}_2, \ldots, \mathbf{r}_N) = \phi_n(\mathbf{r}_2, \mathbf{r}_1, \ldots, \mathbf{r}_N) \quad \text{etc.} \tag{28.4}$$

and the wave function for the system is totally *symmetric*. In this case one speaks of "Bose" particles. One example is the pion, the basic quantum of the nuclear force. Another is atomic ^{4}He which is a composite object consisting of two electrons, two protons, and two neutrons. It is an observed fact, derived in relativistic quantum field theory, that particles with integer spins are bosons.

Or $C^{(1)} = -C^{(2)} = C^{(3)} = -C^{(4)} = \ldots = -C(N!)$ so that

$$\phi_n(\mathbf{r}_1, \mathbf{r}_2, \ldots, \mathbf{r}_N) = -\phi_n(\mathbf{r}_2, \mathbf{r}_1, \ldots, \mathbf{r}_N) \quad \text{etc.} \tag{28.5}$$

and the wavefunction is totally *antisymmetric*. In this case one speaks of "Fermi" particles. In nature, particles with half-integral spins are fermions. Examples are electrons, protons, and neutrons. Another example is atomic ^{3}He, which is again a composite object, but now with two electrons, two protons, and only one neutron. Actually, for the full description of the wave function in these cases, one must specify not only the position vector $\mathbf{r}$ of each particle, but also each particle's spin $s = \pm 1/2$

$$\phi_n(\mathbf{r}_1 s_1, \mathbf{r}_2 s_2, \ldots, \mathbf{r}_N s_N) = -\phi_n(\mathbf{r}_2 s_2, \mathbf{r}_1 s_1, \ldots, \mathbf{r}_N s_N) \quad \text{etc.} \tag{28.6}$$

It is sufficient for our subsequent discussion, however, to consider separately all particles with $s = +1/2$ and all those with $s = -1/2$.

For a single free particle in a state with momentum $\mathbf{p}$, the wavefunction takes the form

$$\phi_{\mathbf{p}}(\mathbf{r}) = \frac{1}{\sqrt{L^3}} e^{i\mathbf{p}\cdot\mathbf{r}/\hbar} \tag{28.7}$$

where

$$p_i = \frac{2\pi\hbar}{L} n_i \quad ; \ n_i = 0, \pm 1, \pm 2, \ldots \tag{28.8}$$

$$i = x, y, z$$

If particle 1 is in the state $\mathbf{p}_1$, particle 2 is in the state $\mathbf{p}_2$, etc., then the total wavefunction for the system takes the form

$$\phi_{\mathbf{p}_1,\mathbf{p}_2\ldots,\mathbf{p}_N}(\mathbf{r}_1,\mathbf{r}_2,\ldots,\mathbf{r}_N) = \prod_{s=1}^{N} \phi_{\mathbf{p}_s}(\mathbf{r}_s) \tag{28.9}$$

and the energy of the system is given by

$$E = \sum_{s=1}^{N} \frac{\mathbf{p}_s^2}{2m} \tag{28.10}$$

The wavefunction must now be symmetrized or antisymmetrized, depending on whether one is describing a system of bosons or of fermions. We now can no longer distinguish the particles, and the state is characterized uniquely not by specifying which particle is in which state, but only by specifying the *occupation number* of each state, irrespective of the particle in the particular state. We write the occupation number n_s as

$$n_s \equiv \text{occupation number}$$
$$\equiv \text{number of particles in the state } s \tag{28.11}$$

With n_s particles in the state $\mathbf{p}_s$, the energy can be rewritten as

$$E_{\{n_s\}} = \sum_{s=1}^{\infty} n_s \frac{\mathbf{p}_s^2}{2m} \tag{28.12}$$

and since the total number of particles is N †

$$N = \sum_{s=1}^{\infty} n_s \tag{28.13}$$

---

† Note that for finite N, most of the n_s will be zero.

Alternatively, if

$$\varepsilon_s \equiv \frac{\mathbf{p}_s^2}{2m} \qquad (28.14)$$

then the energy of the system is expressed as

$$E_{\{n_s\}} = \sum_{s=1}^{\infty} n_s \varepsilon_s \qquad (28.15)$$

With bosons, any values of the occupation numbers are allowed

$$n_s = 0, 1, 2, \ldots, N \qquad ; \text{ bosons} \qquad (28.16)$$

In contrast, it is evident upon antisymmetrizing the wavefunction in Eq.(28.9) that one cannot put two identical fermions into the same single-particle state and obtain a non-vanishing result. This is the *Pauli exclusion principle*, and it implies that the occupation numbers for fermions can only take the values

$$n_s = 0, 1 \qquad ; \text{ fermions} \qquad (28.17)$$

A state n of the entire system is now specified by giving the occupation numbers

$$n \equiv \{n_s\} = \{n_1, n_2 \ldots, n_\infty\} \qquad (28.18)$$

In quantum mechanics with identical particles, there is only *one* state of the entire system corresponding to this set of occupation numbers. Classically, with distinguishable particles there are correspondingly $N!/\prod_s n_s!$ different states.

The quantum partition function for a system of identical particles is given by

$$Z_N(\beta) = \sum_{\{n_s\}}' e^{-\beta E_{\{n_s\}}} = \sum_{\{n_s\}}' e^{-\beta \sum_{s=1}^{\infty} n_s \varepsilon_s} \qquad (28.19)$$

where the sum goes over all sets of occupation numbers satisfying the condition

$$\sum_{s=1}^{\infty} n_s = N \qquad (28.20)$$

The prime in the summation in Eq.(28.19) indicates that the sum is *constrained* by the condition in Eq.(28.20). This restriction is inconvenient; it makes the summations difficult to evaluate. If one could sum over each of the occupation numbers n_s independently, it would be much easier to evaluate the partition function. In order to accomplish this, we turn to a discussion of the grand canonical ensemble.

29. The Grand Canonical Ensemble

The constraint in Eq.(28.20) on the summation in Eq.(28.19) can be removed if one assumes the number of particles to be a *dynamical variable*. To understand how this can be accomplished, we first briefly review the analysis that led to the canonical ensemble. It was first argued from dynamics that for stationary probabilities, one must have an *ensemble* of systems distributed in total energy E according to

$$P = P(E) \tag{29.1}$$

$P(E)$ now has the interpretation as the probability of finding a system with the energy E. It was then argued, in Secs. 9 and 23, that if two systems A and B are in thermodynamic equilibrium with a heat bath and with each other as illustrated in Figure 23.1, then the probability for the combined system C must be the product of individual probabilities

$$P_C(E_A + E_B) = P_A(E_A) \cdot P_B(E_B) \tag{29.2}$$

This relation implies that the probability must have the *form*

$$P(E) = Ce^{-\beta E} \tag{29.3}$$

where β is a constant characterizing thermodynamic equilibrium. If the allowed energies of a system are denoted by $\{E_n\} = \{E_1, E_2, \ldots, E_n, \ldots, E_\infty\}$, then the probability of finding a particular value E_n in the canonical ensemble is

$$P(E_n) = Ce^{-\beta E_n} \tag{29.4}$$

Since the total probability must be unity

$$\sum_n P(E_n) = 1 \tag{29.5}$$

the constant C in Eq.(29.4) can be determined to be

$$C^{-1} = \sum_n e^{-\beta E_n} \equiv \mathcal{Z}_N(\beta) \tag{29.6}$$

This is just the partition function for the canonical ensemble, whose general quantum definition is given by Eq.(24.3)

$$\mathcal{Z}_N(\beta) = \text{Trace}\,(e^{-\beta H}) \tag{29.7}$$

It was shown in Secs. 5 and 12 that for a very large number N of subsystems, the distribution in energy in the canonical ensemble will be *strongly peaked* about the mean value $\bar{E}$. One can thus sensibly assign the energy $\bar{E}$ to a system in thermodynamic equilibrium.

We proceed to make the total number of subsystems N a dynamical variable by working in analogy to the above. Consider first an ensemble of systems distributed in N according to

$$P = P(N) \tag{29.8}$$

P(N) is then the probability of finding a system with a particular value of N. For simplicity, we limit the present discussion to identical subsystems, but it can easily be extended (Prob. 29.1). Assume next that each system is free to exchange particles with a "particle bath." The condition of *thermodynamic equilbrium* can be imposed exactly as in the previous discussion of the energy. If two systems A and B are in thermodynamic equilibrium with the particle bath and with each other, then the probability for the combined system C must be the product of the individual probabilities

$$P(N_A + N_B) = P(N_A) \cdot P(N_B) \tag{29.9}$$

As above, this implies a form

$$P(N) = C' e^{-\lambda N} \tag{29.10}$$

where λ is a new constant characterizing thermodynamic equilibrium.

Now replace each of the preceding individual systems by a canonical ensemble of systems distributed in energy according to

$$P(N, E) = C'' e^{-\lambda N} e^{-\beta E} \tag{29.11}$$

Gibbs called the collection of canonical ensembles with the total particle number N distributed according to Eq.(29.10) the *Macro (Grand) Canonical Ensemble*. It is evidently an ensemble of ensembles. Since the total probability taken over all values of N and E must again be unity

$$\sum_N \sum_n P(N, E_n) = 1 \tag{29.12}$$

the constant C'' can be determined to be

$$[C'']^{-1} \equiv \mathcal{Z}_G(\lambda, \beta) = \sum_N \sum_n e^{-\lambda N} e^{-\beta E_n} \tag{29.13}$$

This expression defines the *quantum grand partition function*

$$\mathcal{Z}_G(\lambda, \beta) \equiv \sum_N \sum_n e^{-\lambda N} e^{-\beta E_n} = \sum_N e^{-\lambda N} \mathcal{Z}_N(\beta) \qquad (29.14)$$

The general definition of the quantum grand partition function is

$$\mathcal{Z}_G(\lambda, \beta) = \text{Trace} \left(e^{-\lambda N} e^{-\beta H} \right) \qquad (29.15)$$

where Trace now includes a sum over states with any number of particles.

The justification for this procedure is again the observation that for a very large number of subsystems N, the distribution in the grand canonical ensemble will be strongly peaked about the mean value $\bar{N}$. One can thus sensibly assign the particle number $\bar{N}$ to a system in thermodynamic equilibrium. The equilibrium situation in the grand canonical ensemble is illustrated in Figure 29.1.

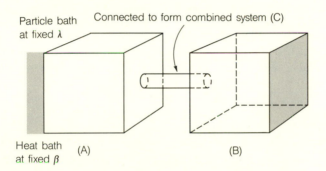

Figure 29.1 Two systems in mutual equilibrium free to exchange particles with a particle bath and energy with a heat bath, and with each other.

We proceed in analogy with the discussion of energy in Secs. 10 and 24 to choose the constant λ such that the average value of $\bar{N}$ will be exactly what we desire for our system

$$\bar{N} = \frac{\sum_N N(e^{-\lambda N} \mathcal{Z}_N(\beta))}{\sum_N e^{-\lambda N} \mathcal{Z}_N(\beta)} \qquad (29.16)$$

Now in the notation of Eq.(28.14)

$$\varepsilon_s = \frac{\mathbf{p}_s^2}{2m} \qquad (29.17)$$

the partition function we were attempting to evaluate in Eq.(28.19) takes the form

$$\mathcal{Z}_N(\beta) = \sideset{}{'}\sum_{\{n_s\}} e^{-\beta \sum_{s=1}^{\infty} n_s \varepsilon_s} \tag{29.18}$$

In fact, the energy ε_s in Eq.(29.17) can be generalized from non-interacting particles to include a one-body potential; then ε_s is just the energy level of the individual particle determined by the one-body hamiltonian. With the aid of Eq.(28.20) one has

$$e^{-\lambda N} = e^{-\lambda \sum_{s=1}^{\infty} n_s} \tag{29.19}$$

The quantum grand partition function introduced in Eq.(29.14) then takes the form†

$$\mathcal{Z}_G(\beta, \lambda) = \sum_{\{n_s\}} e^{-\sum_{s=1}^{\infty} n_s (\beta \varepsilon_s + \lambda)} \tag{29.20}$$

where the sum now goes over all sets of $\{n_s\} = \{n_1, n_2, \ldots, n_\infty\}$ in a fashion *unconstrained* except for the statistics of the particles! We have now accomplished the goal set out at the beginning of this section.

The constant λ is a pure number and it is convenient to take out a factor of β, so we define

$$-\lambda \equiv \beta \mu \tag{29.21}$$

where μ is the "chemical potential." It has units of energy, and it will turn out to be, in fact, the energy required to add a particle to the system at constant volume and entropy. We determine λ in Eq.(29.16) by requiring that N be correct, at a given T; the chemical potential μ is then determined.

Thus one can write

$$\mathcal{Z}_G(\beta, \lambda) = \sum_{\{n_s\}} e^{-\beta \sum_{s=1}^{\infty} n_s (\varepsilon_s - \mu)} \tag{29.22}$$

where the sum on the set of occupation numbers $\{n_s\}$ is now *unrestricted* except for the particle statistics. Equation (29.16) then takes the form

$$\bar{N} = \frac{\sum_{\{n_s\}} \left(\sum_{s=1}^{\infty} n_s \right) e^{-\beta \sum_{s=1}^{\infty} (\varepsilon_s - \mu) n_s}}{\mathcal{Z}_G(\beta, \lambda)} \tag{29.23}$$

---

† There is an additional suppressed dependence on the volume $V = L^3$ which enters through the single-particle energies ε_s.

so that

$$\bar{N} = -\left(\frac{\partial ln\mathcal{Z}_G}{\partial \lambda}\right)_\beta = \frac{1}{\beta}\left(\frac{\partial ln\mathcal{Z}_G}{\partial \mu}\right)_\beta \tag{29.24}$$

Note that for a given set of single-particle energies ε_s, there will be two different answers for the quantum grand partition function and for N: one for bosons and one for fermions.

30. Fermi Statistics

In the case of Fermi statistics, the quantum grand partition function in Eq.(29.22) takes the form

$$\mathcal{Z}_G(\beta, \lambda) = \sum_{\{n_s\}} e^{-\beta \sum_{s=1}^{\infty} n_s(\varepsilon_s - \mu)} = \prod_{s=1}^{\infty}\left(\sum_{n_s} e^{-\beta n_s(\varepsilon_s - \mu)}\right) \tag{30.1}$$

where the sum on the individual n_s is unrestricted except by the statistics

$$n_s = 0, 1 \qquad ; \text{ fermions} \tag{30.2}$$

Note the crucial factoring of the sum in Eq.(30.1) into the product of individual sums, one for each mode; the occupation number of the sth mode is now denoted by n_s. These sums are immediately evaluated with the aid of Eq.(30.2) to yield

$$\mathcal{Z}_G(\beta, \lambda) = \prod_{s=1}^{\infty}\left[1 + e^{-\beta(\varepsilon_s - \mu)}\right] \tag{30.3}$$

Since the logarithm of a product is the sum of the logarithms, one finds

$$ln\mathcal{Z}_G(\beta, \lambda) = \sum_{s=1}^{\infty} ln\left[1 + e^{-\beta(\varepsilon_s - \mu)}\right] \tag{30.4}$$

The (mean) energy of a system is obtained from the same expression we had for the canonical ensemble in Eq.(24.4)

$$\bar{E} = -\left(\frac{\partial ln\mathcal{Z}_G}{\partial \beta}\right)_{\lambda,V} \tag{30.5}$$

and the (mean) number of particles follows from Eq.(29.24)

$$\bar{N} = -\left(\frac{\partial ln\mathcal{Z}_G}{\partial \lambda}\right)_{\beta,V} = \frac{1}{\beta}\left(\frac{\partial ln\mathcal{Z}_G}{\partial \mu}\right)_{\beta,V} \tag{30.6}$$

Direct evaluation yields

$$\bar{N} = \frac{1}{\beta} \sum_{s=1}^{\infty} \frac{\beta e^{-\beta(\varepsilon_s - \mu)}}{1 + e^{-\beta(\varepsilon_s - \mu)}} \tag{30.7}$$

which gives

$$\bar{N} = \sum_{s=1}^{\infty} \frac{1}{e^{\beta(\varepsilon_s - \mu)} + 1} \tag{30.8}$$

The individual terms in this expression represent the mean number of particles in each of the states; thus we define

$$\overline{n_s} \equiv \frac{1}{e^{\beta(\varepsilon_s - \mu)} + 1} \tag{30.9}$$

This distribution is said to describe *Fermi-Dirac statistics*.

We have made a basic approximation in this development. We are looking at our system as if it had a variable number of particles, when, in fact, the systems to which we shall apply this analysis have a *fixed* number of particles. We hope that the contributing values of the number of particles taken this way will be very close to the *real* value and that fluctuations in the particle number will be unimportant. It is possible, of course, to arrange real physical systems where particles can be exchanged with particle baths at fixed chemical potential, for example, in chemical or biochemical processes where molecules are transported through semi-permeable membranes.

Ideal Gas. Consider a non-interacting quantum gas of fermions obeying periodic boundary conditions in a large box of volume $V = L^3$ as in Figure 30.1.

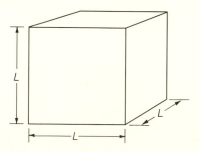

Figure 30.1 Ideal quantum gas in a large box of volume $V = L^3$ obeying periodic boundary conditions.

The allowed values of the momentum are given by Eq.(28.8)

$$p_{x,y,z} = \frac{2\pi\hbar}{L}(n_x, n_y, n_z) \tag{30.10}$$

and the energy is given by

$$\varepsilon_s = \frac{\mathbf{p}_s^2}{2m} \tag{30.11}$$

The density of momentum states is obtained from Eq.(25.15)

$$d\mathbf{n} = \frac{L^3}{h^3}d\mathbf{p} = \frac{V}{h^3}d\mathbf{p} \tag{30.12}$$

In the limit of a large volume, the sum in Eq.(30.8) can be transformed to an integral.

$$\bar{N} = g \cdot \frac{V}{h^3} \int d\mathbf{p} \frac{1}{1 + \exp\left[\beta\left(\mathbf{p}^2/2m \ - \mu\right)\right]} \tag{30.13}$$

Here, consistent with the discussion following Eq.(28.6), a *spin-degeneracy factor* g has been included

$$g \equiv \text{ spin-degeneracy factor} \tag{30.14}$$

This is illustrated in Figure 30.2.

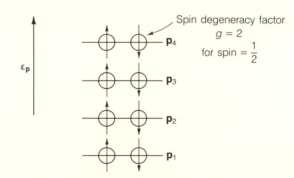

Figure 30.2 Illustration of spin-degeneracy factor g for ideal quantum gas.

As discussed above, we use Eq.(30.13) to *determine* μ for a given N. The particle density $n = N/V$ follows immediately from Eq.(30.13)

$$n \equiv \frac{N}{V} = g \cdot \frac{1}{h^3} \int d\mathbf{p} \frac{1}{1 + \exp\left[\beta\left(\mathbf{p}^2/2m - \mu\right)\right]} \tag{30.15}$$

where we here write $\bar{N} \equiv N$. The integration over angles can be performed with the result that

$$n = g \cdot \frac{4\pi}{h^3} \int_0^\infty p^2 dp \frac{1}{1 + \exp\left[\beta\left(p^2/2m - \mu\right)\right]} \tag{30.16}$$

In differential form, the particle density is

$$dn = g \cdot \frac{1}{h^3} \frac{1}{1 + \exp\left[\beta\left(\mathbf{p}^2/2m - \mu\right)\right]} d\mathbf{p} \tag{30.17}$$

Consider now two limiting cases of these results:

Case 1. $\exp(-\beta\mu) >> 1$

In this case, Eq.(30.17) takes the form

$$dn = e^{\beta\mu} \cdot \frac{g}{h^3} e^{-\beta\mathbf{p}^2/2m} d\mathbf{p} \tag{30.18}$$

which is recognized as the *classical Maxwell-Boltzmann distribution* of Eqs.(13.39-40). The total particle density is obtained by integration

$$n = e^{\beta\mu} \cdot \frac{g}{h^3} \int_0^\infty e^{-\beta\mathbf{p}^2/2m} d\mathbf{p} \tag{30.19}$$

The integral is evaluated in Eqs.(25.16-17)

$$n = e^{\beta\mu} \cdot \frac{g}{h^3} (2\pi m k_B T)^{3/2} \tag{30.20}$$

This relation may now be solved for $\exp(-\beta\mu)$

$$e^{-\beta\mu} = \frac{g}{nh^3} (2\pi m k_B T)^{3/2} \tag{30.21}$$

and the inequality

$$e^{-\beta\mu} >> 1 \tag{30.22}$$

leads to the following condition on the temperature

$$T >> T_{QM}^* \tag{30.23}$$

where

$$T^*_{QM} \equiv \frac{h^2 n^{2/3}}{g^{2/3}(2\pi m k_B)} \tag{30.24}$$

If this condition is met, we have the same results as in the classical treatment of the ideal gas, for Eqs.(30.18) and (30.21) can be combined to yield

$$\frac{dn}{n} = \frac{e^{-\beta \mathbf{p}^2/2m}}{(2\pi m k_B T)^{3/2}} d\mathbf{p} \tag{30.25}$$

which is precisely Eq.(13.39).   The condition in Eq.(30.22) can be rewritten with the aid of Eq.(30.21) as

$$g^{1/3}(2\pi m k_B T)^{1/2} >> h n^{1/3} \tag{30.26}$$

but for a classical ideal gas

$$\frac{1}{2m}\overline{\mathbf{p}^2} = \frac{3}{2}k_B T \tag{30.27}$$

So, to this order of magnitude,

$$\bar{p} >> h n^{1/3} \tag{30.28}$$

This condition can be written as

$$\bar{\lambda} \equiv h/\bar{p} << n^{-1/3} \tag{30.29}$$

Since

$$n^{-1/3} \cong \text{ mean distance between particles} \tag{30.30}$$

The relation in Eq.(30.29) states that one will be in the classical limit for the ideal gas provided the mean *DeBroglie wavelength* $\bar{\lambda} = h/\bar{p}$ is small compared to the mean distance between particles in the gas. In this case, one can build up wave packets for the individual constituents of the system which do not overlap. It is then possible to *distinguish* the particles by their position in space, and the quantum identical particle treatment is of no importance. If the wave packets do not overlap, then one can distinguish the particles.

The temperature in Eq.(30.24) is that at which the spatial extent of the quantum mechanical localization of the particle is of the order of the interparticle spacing. Typical values for ^{3}He gas are

$$m_{3_{He}} \cong 5 \times 10^{-24} g \qquad\qquad ; h = 6.63 \times 10^{-27} \text{erg sec}$$

$$n \cong \frac{6 \times 10^{23}}{20 \times 10^3 \text{cm}^3} = 3 \times 10^{19}/\text{cm}^3 \quad ; k_B = 1.38 \times 10^{-16} \text{erg}(^\circ K)^{-1}$$

$$\tag{30.31}$$

which yields

$$T^*_{QM} \cong 6 \times 10^{-2} \, ^\circ K \tag{30.32}$$

and hence the classical treatment holds at almost all temperatures. On the other hand, typical values for conducting electrons in a metal are

$$m_e = 9.11 \times 10^{-28} g$$

$$n_{cu} \cong 2 \times \frac{8.96 \ g/cm^3}{63.54 \ g/mole} \times 6.02 \times 10^{23}/mole = 1.70 \times 10^{23}/cm^3 \tag{30.33}$$

and hence the condition in Eq.(23.23) under which classical statistics holds is

$$T \gg T^*_{QM} \cong 1.08 \times 10^5 \, ^\circ K \tag{30.34}$$

which is a very high temperature indeed.

Case 2. $\exp(-\beta\mu) \ll 1$

In this second limiting case, it is evident that the chemical potential μ must be positive, so we define

$$\mu \equiv \varepsilon_0 \tag{30.35}$$

The basic inequality that controls this limiting case then becomes

$$e^{\beta\varepsilon_0} \gg 1 \tag{30.36}$$

or

$$\varepsilon_0 \gg k_B T \tag{30.37}$$

Thus, in this case, one is discussing the *low-temperature limit* of the quantum statistics of an ideal Fermi gas. The average occupation number for the sth state in Eq.(30.9) becomes

$$\bar{n}_s = \frac{1}{\exp\left[\beta\left(\mathbf{p}_s^2/2m - \varepsilon_0\right)\right] + 1} \tag{30.38}$$

where the single-particle energy is defined through Eq.(28.14). This Fermi distribution is sketched in Figure 30.3.

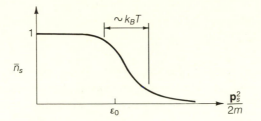

Figure 30.3 Average occupation number of the sth state with energy $\varepsilon_s = \mathbf{p}_s^2/2m$ as defined in Eq.(30.38) for an ideal Fermi gas at low temperature

At $T = 0$ this distribution becomes a unit step function with a step at $\varepsilon_s = \varepsilon_0$, where ε_0 is called the *Fermi energy*. The *Fermi momentum* p_0 is related to the Fermi energy by

$$\varepsilon_0 \equiv \frac{p_0^2}{2m} \tag{30.39}$$

Since in this zero-temperature limit the levels are occupied with unit probability up to the Fermi momentum, and with zero probability above that value, the total number of particles in Eq.(30.16) takes the form

$$\frac{N}{V} \equiv n = g \cdot \frac{4\pi}{h^3} \int_0^{P_o} p^2 dp = g \cdot \frac{4\pi}{3h^3} p_0^3 \tag{30.40}$$

The inversion of this relation expresses the Fermi momentum in terms of the particle density

$$p_0 = h \cdot \left(\frac{3n}{4\pi g} \right)^{1/3} \tag{30.41}$$

The Fermi energy can then be written as

$$\varepsilon_0 = \frac{p_0^2}{2m} = \left(\frac{3n}{4\pi g} \right)^{2/3} \cdot \frac{h^2}{2m} \tag{30.42}$$

Comparison with the temperaure T_{QM}^* defined in Eq.(30.24) yields

$$\varepsilon_0 \sim k_B T_{QM}^* \tag{30.43}$$

Hence the basic inequality defining the low-temperature limit for an ideal Fermi gas in Eq.(30.37) becomes

$$k_B T \ll \varepsilon_0 \sim k_B T_{QM}^* \qquad (30.44)$$

or

$$T \ll T_{QM}^* \qquad (30.45)$$

In this limit, one speaks of a *degenerate Fermi gas*. In contrast to the situation discussed after Eq.(30.29), the DeBroglie wavelength is now large compared to the interparticle spacing, and we cannot treat the particles as distinguishable.

As an example, consider the conduction electrons in a metal to constitute a free Fermi gas. Typical values are then [compare Eq.(30.33)]

$$n \cong 10^{23}/cm^3$$
$$m_e \cong 10^{-27} gm \qquad (30.46)$$

Equation (30.24) and the constants in Eq.(30.31) then yield

$$T_{QM}^* = 6.9 \times 10^4 \, °K \qquad (30.47)$$

Therefore when one is concerned with a metal at room temperature, the system constitutes a *degenerate* Fermi gas. Now this is interesting, because if one has an ideal gas of electrons, then in addition to the T^3 Debye law contribution to the specific heat in Eq.(27.30b) coming from the lattice vibrations, one might expect the additional, constant contribution $c_v = 3R/2$ of Eq.(25.20) arising from the translational motion of the electrons. In fact, there is *no such constant term*. The explanation lies in the fact that the gas is degenerate and its properties are governed by Fermi statistics. Let us make an estimate of the actual electronic contribution to the specific heat. If one goes from the temperature $T = 0$ to a temperature $T > 0$ slightly above zero, then some of the electrons in the Fermi gas will be thermally excited from states just below the Fermi momentum to states just above the Fermi momentum. The distribution is, in fact, given by Eq.(30.38), and the situation is sketched in Figure 30.4.

Let us ask how many of the electrons can increase their energy by an amount $\sim k_B T$. Consider first the volume of the layer in momentum space that is thermally excited. The total number of particles is given by Eq.(30.40)

$$N = g \cdot \frac{4\pi V}{3h^3} p_0^3 \qquad (30.48)$$

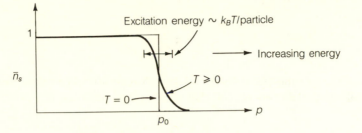

Figure 30.4 Thermal excitation of a degenerate Fermi gas at low temperature.

The differential of this relation then yields

$$\Delta N = g \cdot \frac{4\pi V}{h^3} p_0^2 \Delta p \qquad (30.49)$$

and so

$$\frac{\Delta N}{N} = 3\frac{\Delta p}{p_0} \qquad (30.50)$$

The Fermi energy is given by

$$\varepsilon_0 = \frac{p_0^2}{2m} \qquad (30.51)$$

The differential of the energy relation evaluated at the Fermi energy yields

$$\Delta\varepsilon = \frac{2p_0}{2m}\Delta p \qquad (30.52)$$

and hence

$$\frac{\Delta\varepsilon}{\varepsilon_0} = 2\frac{\Delta p}{p_0} \qquad (30.53)$$

A combination of Eqs.(30.50) and (30.53) gives

$$\frac{\Delta N}{N} = 3\frac{\Delta p}{p_0} = \frac{3}{2}\frac{\Delta\varepsilon}{\varepsilon_0} \qquad (30.54)$$

The introduction of Eq.(30.43) then implies that the number of electrons close to the Fermi surface which increase their energy by an amount $\Delta\varepsilon \sim k_B T$ is given approximately by

$$\frac{\Delta N}{N} \sim \frac{k_B T}{k_B T^*_{QM}} \tag{30.55}$$

If E_0 is the ground-state energy of the degenerate Fermi gas, then the thermal exctitation energy of that gas at low temperature is approximately

$$E - E_0 \sim (k_B T)\Delta N \tag{30.56}$$

which yields

$$E - E_0 \sim k_B T \left(\frac{NT}{T^*_{QM}}\right) \sim Nk_B \frac{T^2}{T^*_{QM}} \tag{30.57}$$

The constant-volume heat capacity is therefore

$$C_V = \left[\frac{\partial(E - E_0)}{\partial T}\right]_V \sim Nk_B \left(\frac{T}{T^*_{QM}}\right) \tag{30.58}$$

This result is *linear* in the temperature. Furthermore, at room temperature where

$$\frac{T}{T^*_{QM}} << 1 \tag{30.59}$$

the actual contribution of this term to the specific heat of the metal is small. At low temperature where the Debye T^3 contribution from the phonons is insignificant, this linear term coming from the translational motion of the electrons can be measured.

The low-temperature heat capacity of an ideal Fermi gas estimated in Eq.(30.58) can, in fact, be calculated analytically (Prob. 30.3). The result is†

$$c_v = \frac{\pi^2}{2} Nk_B \left(\frac{k_B T}{\varepsilon_0}\right)$$
$$= \frac{\pi}{2}\left(\frac{4\pi}{3}\right)^{2/3} Nk_B \left(\frac{T}{T^*_{QM}}\right) \tag{30.60}$$

---

† See A. L. Fetter and J. D. Walecka, *Quantum Theory of Many-Particle Systems, op. cit.* p.48.

31. Bose Statistics

In the case of Bose statistics, the quantum grand partition function takes the form

$$\mathcal{Z}_G(\beta, \lambda) = \sum_{\{n_s\}} e^{-\beta \sum_{s=1}^{\infty} n_s(\varepsilon_s - \mu)} = \prod_{s=1}^{\infty} \left(\sum_{n_s} e^{-\beta n_s(\varepsilon_s - \mu)} \right) \quad (31.1)$$

where the sum on the individual n_s is again unrestricted except by the statistics. In the case of bosons, however, the allowed occupation numbers are

$$n_s = 0, 1, 2, \ldots, \infty \quad (31.2)$$

The sums in Eq.(31.1) can be immediately performed, they are just geometric series, and one finds

$$\mathcal{Z}_G(\beta, \lambda) = \prod_{s=1}^{\infty} \left(\frac{1}{1 - e^{-\beta(\varepsilon_s - \mu)}} \right) \quad (31.3)$$

The logarithm of this expression is

$$ln\mathcal{Z}_G(\beta, \lambda) = -\sum_{s=1}^{\infty} ln \left(1 - e^{-\beta(\varepsilon_s - \mu)} \right) \quad (31.4)$$

and the mean number of particles follows from Eq.(29.24)

$$\bar{N} = \frac{1}{\beta} \left(\frac{\partial ln\mathcal{Z}_G}{\partial \mu} \right)_{\beta, V} \quad (31.5)$$

Direct evaluation yields

$$\bar{N} = \frac{1}{\beta} \sum_{s=1}^{\infty} \frac{\beta e^{-\beta(\varepsilon_s - \mu)}}{1 - e^{-\beta(\varepsilon_s - \mu)}} \quad (31.6)$$

which gives

$$\bar{N} = \sum_{s=1}^{\infty} \frac{1}{e^{\beta(\varepsilon_s - \mu)} - 1} \quad (31.7)$$

The individual terms in this expression again represent the mean number of particles in each of the states; thus we define

$$\bar{n}_s \equiv \frac{1}{e^{\beta(\varepsilon_s - \mu)} - 1} \quad (31.8)$$

This distribution is said to describe *Bose-Einstein statistics*. Note particularly the -1 in the denominator, in contrast to the Fermi-Dirac distribution of Eq.(30.9) where $a+1$ appears. At the present stage of the argument, this is the only difference between the two statistics.

Ideal Gas. Consider a non-interacting quantum gas of bosons obeying periodic boundary conditions in a large box of volume $V = L^3$ as in Figure 30.1. The arguments of Eqs.(30.10) to (30.17) may be taken over verbatim, the only difference being the sign in the denominator of the distribution function. Thus we can write

$$dn = g \cdot \frac{4\pi}{h^3} \frac{1}{e^{\beta(p^2/2m - \mu)} - 1} p^2 dp \tag{31.9}$$

and

$$n = \frac{N}{V} = g \cdot \frac{4\pi}{h^3} \int_0^\infty p^2 dp \frac{1}{e^{\beta(p^2/2m - \mu)} - 1} \tag{31.10}$$

where we here and henceforth write $\bar{N} \equiv N$. Consider now the same two limiting cases of these results discussed in the preceding section:

Case 1. $\exp(-\beta\mu) \gg 1$

In this case, the -1 can be neglected in the denominator of Eq.(31.8) and one has from Eq.(31.9)

$$dn = e^{\beta\mu} \cdot g \cdot \frac{4\pi}{h^3} e^{-\beta \mathbf{p}^2/2m} \cdot p^2 dp \tag{31.11}$$

This is the same result as in Eq.(30.18); it clearly does not matter whether one starts from Bose-Einstein or Fermi-Dirac statistics in arriving at this result. The answer is the same in the limit

$$e^{-\beta\mu} \gg 1 \tag{31.12}$$

for then the additional ± 1 in the denominator is of no significance and the identity of the particles plays no role; the reason is discussed in the paragraph following Eq.(30.30). In the limit of Eq.(30.23)

$$T \gg T_{QM}^* \equiv \frac{h^2 n^{2/3}}{g^{2/3}(2\pi m k_B)} \tag{31.13}$$

one recovers classical statistical mechanics.

The first physical system to which one might attempt to apply this analysis is ^{4}He, whose atoms are bosons. At low pressures, ^{4}He remains a liquid down to absolute zero.[†] The density of this liquid is

$$\rho_{^4He} \cong 0.145 g/cm^3 \tag{31.14}$$

[†] For a discussion of superfluid ^{4}He, see Fetter and Walecka, *Quantum Theory of Many-Particle Systems, op. cit.*, p.479f.

and hence

$$n_{^4He} \cong \frac{0.145g/cm^3}{4g/\text{mole}} \times 6.02 \times 10^{23} \text{atoms/mole} \cong 2.2 \times 10^{22}/cm^3$$

$$m_{^4He} \cong 6.7 \times 10^{-24}g$$

(31.15)

As in Eqs.(30.31-32), the temperature T_{QM}^* is given in this case by

$$T_{QM}^* \cong 5.9 \,^\circ K \qquad (31.16)$$

Case 2. $\exp(-\beta\mu) \cong 1$

It is evident from Eq.(31.8) that the chemical potential for an ideal Bose gas can never become positive, for then there would exist states with negative occupation numbers. Thus one can never go to the limit $\exp(-\beta\mu) << 1$ as in the case of fermions. The closest one can get to this limiting case with bosons is $\exp(-\beta\mu) \cong 1$, which implies

$$\beta\mu \cong 0 \qquad (30.17)$$

In this case, the particle density in Eq.(31.10) is given by

$$n = g \cdot \frac{4\pi}{h^3} \int_0^\infty p^2 dp \frac{1}{e^{\beta_c p^2/2m} - 1} \qquad (31.18)$$

This is a definite integral which is directly reduced to†

$$n = g \cdot \frac{4\pi}{h^3} \left(\frac{2m}{\beta_c}\right)^{3/2} \frac{1}{2} \int_0^\infty \frac{\sqrt{x}dx}{e^x - 1}$$

$$= g \cdot \frac{2\pi}{h^3} (2mk_B T_c)^{3/2} \cdot (2.315\ldots) \qquad (31.19)$$

where T_c is now defined to be that temperature at which the chemical potential reaches zero from below. Solution for T_c gives

$$T_c = \frac{h^2}{2mk_B} \left[\frac{n}{2\pi g(2.315)}\right]^{2/3} \qquad (31.20)$$

which implies

$$T_c = 0.527 \, T_{QM}^* \qquad (31.21)$$

---

† See Fetter and Walecka, *Quantum Theory of Many-Particle Systems, op. cit.*

If $T \leq T_c$ we run into problems. It is evident from Eq.(31.19) that under this condition

$$\frac{n(T)}{n(T_c)} = \left(\frac{T}{T_c}\right)^{3/2} \tag{31.22}$$

and one obtains less than the total density by integration over the occupation numbers. What has happened to the remaining particles?

$$V[n(T_c) - n(T)] \equiv N - N' \equiv \Delta N \tag{31.23}$$

The explanation lies in the fact that we have approximated a sum by an integral. If indeed $\mu = 0$, then the occupation number for the one state with $\mathbf{p} = 0$ in Eq.(31.8) is *infinite*, and this contribution is not recovered from the integral in Eq.(31.18). Therefore one must really take the chemical potential to lie just slightly below zero and write

$$-\beta\mu \equiv \kappa \cong 0 \tag{31.24}$$

Then we get a finite occupation of the state with $\mathbf{p} = 0$ and the remaining contribution to the particle number can be obtained by an integration over the remaining occupation numbers

$$N = g \cdot \frac{1}{e^{\kappa} - 1} + 4\pi V \cdot \frac{g}{h^3} \int_0^{\infty} p^2 dp \frac{1}{e^{\beta p^2/2m} - 1} \tag{31.25}$$

Now as $T \longrightarrow 0$, the integral vanishes as in Eq.(31.22) and we get

$$N = g \cdot \frac{1}{\kappa} \qquad ; \; T = 0 \tag{31.26}$$

If one plots the particle distribution in momentum space as in Figure 31.1, a finite fraction of the particles occupy the single zero-energy state; one has *Bose-Einstein condensation* into the state with $\mathbf{p} = 0$.

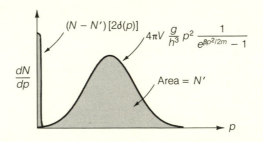

Figure 31.1 Distribution of the particles in momentum space for an ideal Bose gas below the transition temperature T_c. {Note: $\int_0^{\infty} \delta(p)dp = 1/2$.}

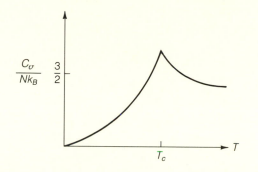

Figure 31.2 Constant-volume heat capacity C_V of an ideal Bose gas.

The constant-volume heat capacity of an ideal Bose gas can be calculated analytically (Prob. 31.3).[†] The result is sketched in Figure 31.2. There is a phase transition and a finite discontinuity in the specific heat at the temperature T_c.

It is an experimental fact that when liquid ^{4}He in contact with its vapor is cooled below the temperature T_λ, where

$$T_\lambda = 2.17\,^\circ K \tag{31.27}$$

there is a phase transition to a new phase with remarkable superfluid properties.[‡] It is intriguing that the insertion of Eq.(31.16) into Eq.(31.21) gives a value $T_c \cong 3.11\,^\circ K$ close to the observed T_λ of liquid ^{4}He.

[†] See Fetter and Walecka, *Quantum Theory of Many-Particle Systems, op. cit.*, p.44.
[‡] *ibid.* p. 479f.

Appendixes

Appendix A

Canonical Transformations and Poisson Brackets(*)

Since Hamilton's equations determine the time-dependence of the set (p, q) of dynamical variables, they likewise determine that of any function $\phi(p, q)$ of these variables. Indeed, from

$$\frac{d\phi}{dt} = \sum_{i=1}^{n} \left(\frac{\partial\phi}{\partial p_i} \frac{dp_i}{dt} + \frac{\partial\phi}{\partial q_i} \frac{dq_i}{dt} \right) \tag{A.1}$$

one obtains in view of Eqs.(1.3)

$$\frac{d\phi}{dt} = \sum_{i=1}^{n} \left(-\frac{\partial\phi}{\partial p_i} \frac{\partial H}{\partial q_i} + \frac{\partial\phi}{\partial q_i} \frac{\partial H}{\partial p_i} \right) \tag{A.2}$$

Considering two arbitrary functions $A(p, q)$ and $B(p, q)$, the *Poisson Bracket* is defined by

$$[A, B]_{P.B.} = \sum_{i=1}^{n} \left(\frac{\partial A}{\partial q_i} \frac{\partial B}{\partial p_i} - \frac{\partial A}{\partial p_i} \frac{\partial B}{\partial q_i} \right) \tag{A.3}$$

so that

$$[B, A]_{P.B.} = -[A, B]_{P.B.} \tag{A.4}$$

Equation (A.2) can then be written in the form

$$\frac{d\phi}{dt} = -[H, \phi]_{P.B.} \tag{A.5}$$

In view of the closely parallel role of momenta and coordinates, it is indicated to replace the pair p_i and q_i with $i = 1, 2, \ldots, n$ by variables x_κ with $\kappa = 1, 2, \ldots 2n$ such that

$$\begin{aligned} p_i &= x_{2i-1} \\ q_i &= x_{2i} \quad ; \qquad i = 1, 2, \ldots, n \end{aligned} \tag{A.6}$$

Functions of the set (p, q) of pairs p_i and q_i are thus to be regarded as functions of the set (x) of variables x_κ.

With Greek indices to stand for any number from 1 to $2n$, this notation calls for the introduction of the matrix

$$\varepsilon_{\kappa\lambda} = \sum_{i=1}^{n} \varepsilon_{\kappa\lambda}^{i} \tag{A.7}$$

where

$$\varepsilon^{i}_{\kappa\lambda} = \delta_{\kappa,2i-1}\,\delta_{\lambda,2i} - \delta_{\kappa,2i}\,\delta_{\lambda,2i-1} \tag{A.8}$$

and where the δ-symbol has the usual significance

$$\delta_{a,b} = \begin{cases} 1 & \text{for } a = b \\ 0 & \text{for } a \neq b \end{cases} \tag{A.9}$$

Since $\varepsilon^{i}_{\kappa\lambda}$ changes sign upon an interchange of κ and λ, the same holds for $\varepsilon_{\kappa\lambda}$, so that

$$\varepsilon_{\lambda\kappa} = -\varepsilon_{\kappa\lambda} \tag{A.10}$$

The matrix elements of $\varepsilon_{\kappa\lambda}$ are further seen from Eqs.(A.7) and (A.8) to have values

$$\varepsilon_{\kappa\lambda} = \begin{cases} 1 & \text{for } \lambda \text{ even and } \kappa = \lambda - 1 \\ -1 & \text{for } \lambda \text{ odd and } \kappa = \lambda + 1 \\ 0 & \text{otherwise} \end{cases} \tag{A.11}$$

The determinant of this matrix is therefore the nth power of the determinant

$$\begin{vmatrix} 0 & 1 \\ -1 & 0 \end{vmatrix} = 1$$

and, hence has the value

$$|\varepsilon| = 1 \tag{A.12}$$

In the customary convention it shall be understood from now on that the same Greek index appearing twice in an expression implies a summation over that index from 1 to $2n$. The definition of the Poisson Bracket by Equation (A.3) can then be written in the form

$$[A, B]_{P.B.} = -\frac{\partial A}{\partial x_{\kappa}}\,\varepsilon_{\kappa\lambda}\,\frac{\partial B}{\partial x_{\lambda}} \tag{A.13}$$

Indeed, with $\varepsilon_{\kappa\lambda}$ from Eqs.(A.7) and (A.8), one obtains on the right side

$$[A, B]_{P.B.} = -\sum_{i=1}^{n}\left(\frac{\partial A}{\partial x_{2i-1}}\frac{\partial B}{\partial x_{2i}} - \frac{\partial A}{\partial x_{2i}}\frac{\partial B}{\partial x_{2i-1}}\right) \tag{A.14}$$

which, in view of Eq.(A.6), stands in this notation for the expression on the right side of Eq.(A.3). Writing in Equation (A.13) λ' for λ, one has thus from Eq.(A.5)

$$\frac{d\phi}{dt} = \frac{\partial H}{\partial x_{\kappa}}\,\varepsilon_{\kappa\lambda'}\,\frac{\partial \phi}{\partial x_{\lambda'}} \tag{A.15}$$

Choosing in particular $\phi = x_\lambda$ and noting that $\partial x_\lambda / \partial x_{\lambda'} = \delta_{\lambda,\lambda'}$, one has therefore

$$\frac{dx_\lambda}{dt} = \frac{\partial H}{\partial x_\kappa} \varepsilon_{\kappa\lambda} \tag{A.16}$$

as another formulation of Hamilton's equations. In fact, with $\varepsilon_{\kappa\lambda}$ from Eq.(A.11), the implied summation over κ is seen for $\lambda = 2i - 1$ and $\lambda = 2i$, respectively, to yield

$$\frac{dx_{2i-1}}{dt} = -\frac{\partial H}{\partial x_{2i}}$$
$$\frac{dx_{2i}}{dt} = \frac{\partial H}{\partial x_{2i-1}} \quad ; \ i = 1, \ 2, \ \ldots, \ n \tag{A.17}$$

thus verifying the equivalency with Eq.(1.3) in the notation of Eq.(A.6).

Starting from the description of a given mechanical system by the set (p, q) of variables p_i and q_i with $i = 1, \ 2, \ \ldots, \ n$, one may go over to a new set (P, Q) of variables P_k and Q_k with $k = 1, \ 2, \ \ldots, \ n$, defined as functions of the former through

$$P_k = P_k(p, q) \quad ; \quad Q_k = Q_k(p, q) \tag{A.18}$$

or the inverse relations

$$p_i = p_i(P, Q) \quad ; \quad q_i = q_i(P, Q) \tag{A.19}$$

The set (P, Q) shall be replaced by a set (X) of variables X_μ ($\mu = 1, \ 2, \ \ldots, \ 2n$) such that in analogy to Equation (A.6)

$$P_k = X_{2k-1} \quad ; \quad Q_k = X_{2k} \tag{A.20}$$

Equivalent to the transformation of the sets (p, q) and (P, Q), given by Eqs.(A.18) and (A.19), one then deals with the relations,

$$X_\mu = X_\mu(x) \tag{A.21}$$

or

$$x_\kappa = x_\kappa(X) \tag{A.22}$$

to characterize a general transformation from the previous set (x) to the new set (X) of variables or vice versa.

Among all possible transformations one has to consider in particular the restricted class of those called *canonical transformations* under which the formulation of classical mechanics remains unchanged. This restriction can be expressed in a number of ways, chosen according to its manifestation under different aspects. For the present purposes, a canonical transformation shall be defined by the property to lead to the invariance of the Poisson Bracket in the sense that it allows in Eq.(A.13) the formal replacement of the set (x) by the set (X). One thus requires the equality of

$$[A, B]_{(x)} = -\frac{\partial A}{\partial x_\kappa} \varepsilon_{\kappa\lambda} \frac{\partial B}{\partial x_\lambda} \qquad (A.23)$$

and

$$[A, B]_{(X)} = -\frac{\partial A}{\partial X_\mu} \varepsilon_{\mu\nu} \frac{\partial B}{\partial X_\nu} \qquad (A.24)$$

for any two quantities A and B, expressed in their dependence either on the variables of the set (x) or on those of the set (X).† With the latter leading to the former by substitution from Equation (A.21), one has generally

$$\frac{\partial A}{\partial x_\kappa} = \frac{\partial A}{\partial X_\mu} \frac{\partial X_\mu}{\partial x_\kappa} \quad ; \quad \frac{\partial B}{\partial x_\lambda} = \frac{\partial B}{\partial X_\nu} \frac{\partial X_\nu}{\partial x_\lambda} \qquad (A.25)$$

and hence from Eq.(A.23)

$$[A, B]_{(x)} = -\frac{\partial A}{\partial X_\mu} \frac{\partial X_\mu}{\partial x_\kappa} \varepsilon_{k\lambda} \frac{\partial X_\nu}{\partial x_\lambda} \frac{\partial B}{\partial X_\nu} \qquad (A.26)$$

The required agreement with $[A, B]_{(X)}$ from Eq.(A.24) is thus obtained through the condition

$$\frac{\partial X_\mu}{\partial x_\kappa} \varepsilon_{\kappa\lambda} \frac{\partial X_\nu}{\partial x_\lambda} = \varepsilon_{\mu\nu} \qquad (A.27)$$

to be satisfied by a canonical transformation. This condition can also be written in the form

$$[X_\mu, X_\nu]_{(x)} = -\varepsilon_{\mu\nu} \qquad (A.28)$$

with the expression of the Poisson Bracket given in Eq.(A.23). With the expression of Eq.(A.24), however, Eq.(A.28) is seen to merely represent an identity.

The invariance of the Poisson Bracket under a canonical transformation assures that of Hamilton's equations in the sense that their form of Eqs.(A.16) for the variables of the set (x) goes over into the form

$$\frac{dX_\nu}{dt} = \frac{\partial H}{\partial X_\mu} \varepsilon_{\mu\nu} \qquad (A.29)$$

---

† The *explicit* notation of Eq.(6.28) is thus suppressed in this development.

for those of the set (X). Indeed, the former was seen to follow from Eq.(A.5), writing in the notation of Eq.(A.23)

$$\frac{d\phi}{dt} = -[H, \phi]_{(x)} \qquad (A.30)$$

and choosing $\phi = x_\lambda$. Since a canonical transformation allows us to write instead

$$\frac{d\phi}{dt} = -[H, \phi]_{(X)} \qquad (A.31)$$

with the Poisson Bracket given by the expression of Eq.(A.24), the analogous argument leads to Eq.(A.29) upon choosing $\phi = X_\nu$.

Going back to the set of variables (p, q) and (P, Q), the condition of Eq.(A.28) for a canonical transformation shall be reformulated so as to represent the corresponding restriction of a general transformation in the form of Eqs.(A.18) and (A.19). The notation, introduced in Eqs.(A.6) and (A.20) leads to this reformulation by letting the indices μ and ν independently assume even and odd values. In the separate cases of (μ, ν), indicated by $(2k - 1, 2\ell), (2k - 1, 2\ell - 1), (2k, 2\ell)$ and with these pairs of indices inserted for (κ, λ) in Eq.(A.11) one thus arrives at the three relations

$$[P_k, Q_\ell]_{(p,q)} = -\delta_{k\ell} \qquad (A.32)$$

$$[P_k, P_\ell]_{(p,q)} = 0 \qquad (A.33)$$

$$[Q_k, Q_\ell]_{(p,q)} = 0 \qquad (A.34)$$

with the Poisson Bracket to be understood in the original definition of Eq.(A.3). The fourth possibility $(2k, 2\ell - 1)$ yields $[Q_k, P_\ell]_{(p,q)} = \delta_{k\ell}$, so that it merely repeats Eq.(A.32) upon an interchange of k and ℓ and the subsequent interchange P_k and Q_ℓ, accompanied by a change of sign in view of Eq.(A.4).

Similarly, the invariance of the Poisson Bracket is reformulated by verifying that the equality of the expressions in Eqs.(A.23) and (A.24) signifies that of its original form in Eq.(A.3) and the expression

$$[A, B]_{(P,Q)} = \sum_{k=1}^{n} \left[\frac{\partial A}{\partial Q_k} \frac{\partial B}{\partial P_k} - \frac{\partial A}{\partial P_k} \frac{\partial B}{\partial Q_k} \right] \qquad (A.35)$$

respectively. The ensuing invariance of Hamilton's equations is then expressed by the fact that Eqs.(1.2) and (1.3) go over into the equations

$$\frac{dP_k}{dt} = -\frac{\partial H}{\partial Q_k} \qquad (A.36)$$

and

$$\frac{dQ_k}{dt} = \frac{\partial H}{\partial P_k} \tag{A.37}$$

obtained by the reformulation of Eq.(A.29). It is possible, instead, to base this statement of invariance directly upon the condition of a canonical transformation in the form of Eq.(A.32), (A.33), and (A.34), but one loses thereby the previous advantage of a more concise formulation.

Besides many other possibilities, these equations allow an arbitrary transformation

$$Q_k = Q_k(q) \tag{A.38}$$

or

$$q_i = q_i(Q) \tag{A.39}$$

between the sets of coordinates (q) and (Q) above. Since $\partial Q_{k,\ell}/\partial p_i$ vanishes in that case, Eq.(A.34) is here identically satisfied and Eq.(A.32) reduces to

$$\sum_{i=1}^{n} \frac{\partial P_k}{\partial p_i} \frac{\partial Q_\ell}{\partial q_i} = \delta_{k\ell} \tag{A.40}$$

It is satisfied by the relation

$$P_k = \sum_{j=1}^{n} p_j \frac{\partial q_j}{\partial Q_k} \tag{A.41}$$

so that

$$\frac{\partial P_k}{\partial p_i} = \frac{\partial q_i}{\partial Q_k} \tag{A.42}$$

and hence

$$\sum_{i=1}^{n} \frac{\partial q_i}{\partial Q_k} \frac{\partial Q_\ell}{\partial q_i} = \frac{\partial Q_\ell}{\partial Q_k} = \delta_{k\ell} \tag{A.43}$$

Further, with

$$\frac{\partial P_\ell}{\partial q_i} = \sum_{j=1}^{n} p_j \frac{\partial}{\partial q_i} \left[\frac{\partial q_j}{\partial Q_\ell} \right] \tag{A.44}$$

from Eq.(A.41), the Poisson Bracket of Eq.(A.33) represents the difference between the term

$$-\sum_{i=1}^{n} \frac{\partial P_k}{\partial p_i} \frac{\partial P_\ell}{\partial q_i} = -\sum_{j=1}^{n} p_j \sum_{i=1}^{n} \frac{\partial q_i}{\partial Q_k} \frac{\partial}{\partial q_i} \left[\frac{\partial q_j}{\partial Q_\ell} \right] = -\sum_{j=1}^{n} p_j \frac{\partial^2 q_j}{\partial Q_k \partial Q_\ell} \tag{A.45}$$

and the term obtained by interchanging k and ℓ. Since this interchange does not alter the mixed derivative of q_j, the two terms cancel, thus resulting in a vanishing Poisson Bracket as required by Eq.(A.33). It is therefore sufficient to complement any coordinate transformation with the relation of Eq.(A.41) between the momenta to yield a canonical transformation. Equation (A.41) is precisely Eq.(1.49), applications of which are discussed in the text.

Appendix B

General Proof of Liouville's Theorem(*)

Using the notations of Eq.(A.6) for the dynamical variables p_i and q_i, the variables x_κ may be regarded as the coordinates in a $2n$-dimensional space, called the *phase space*. Every set of these variables is thus represented by a point and the quantity

$$d\lambda = \prod_{i=1}^{n} dp_i dq_i \qquad (B.1)$$

or

$$d\lambda = \prod_{\kappa=1}^{2n} dx_\kappa \qquad (B.2)$$

represents the volume element of the phase space. More generally, one may consider $2n$ differential vectors $d\mathbf{x}^\alpha$ and the volume element

$$d\lambda = |dx_\kappa^\alpha| \qquad (B.3)$$

obtained as the determinant of the components $dx_\kappa^\alpha (\alpha, \kappa = 1, 2, \dots, 2n)$ so that Eq.(B.2) refers to the particular choice $dx_\kappa^\alpha = dx_\kappa$ for $\alpha = \kappa$ and $dx_\kappa^\alpha = 0$ for $\alpha \neq \kappa$.

A transformation of variables given by Eq.(A.21) can be interpreted as a mapping of the phase space in the sense that a point with coordinates x_κ goes over into a point with coordinates $X_\mu(x)$. The components of the differential vectors $d\mathbf{x}^\alpha$ thus lead to the components

$$dX_\mu^\alpha = \frac{\partial X_\mu}{\partial x_\kappa} dx_\kappa^\alpha \qquad (B.4)$$

of the corresponding differential vectors $d\mathbf{X}^\alpha$ and, in analogy to Eq.(B.3) to the volume-element

$$d\Lambda = |dX_\mu^\alpha| \qquad (B.5)$$

In view of Eq.(B.4) one obtains from the rules of determinant multiplication

$$d\Lambda = J d\lambda \qquad (B.6)$$

as the relation between corresponding volume-elements where

$$J = \left| \frac{\partial X_\mu}{\partial x_\kappa} \right| \qquad (B.7)$$

is the jacobian determinant of the transformation. It is an important property of the phase space, known as *Liouville's theorem*, that the volume-element is invariant under a canonical transformation so that

$$d\Lambda = d\lambda \qquad (B.8)$$

or, in view of Eq.(B.6), that the jacobian has the value

$$J = 1 \qquad (B.9)$$

Since the volume, contained in any region of the phase space, can be divided into its constituting volume elements, this theorem states that, regardless of a different shape acquired in the mapping by a canonical transformation, the volume of the region remains unchanged.

The complete proof of Eq.(B.9) and, hence, of Liouville's theorem will be presented below. As a less stringent but far more direct conclusion, it shall first be shown that

$$J^2 = 1 \qquad (B.10)$$

Indeed, considering the condition for a canonical transformation formulated in Eq.(A.27), it follows from the equality of the determinants on both sides and from the rule of determinant multiplication that

$$\left|\frac{\partial X_\mu}{\partial x_\kappa}\right| |\varepsilon_{\kappa\lambda}| \left|\frac{\partial X_\nu}{\partial x_\lambda}\right| = |\varepsilon_{\mu\nu}| \qquad (B.11)$$

Since the notation used to label the rows and columns of a determinant is irrelevant to its value, so that

$$\left|\frac{\partial X_\mu}{\partial x_\lambda}\right| = \left|\frac{\partial X_\mu}{\partial x_\kappa}\right| = J \qquad (B.12)$$

and since

$$|\varepsilon_{\mu\nu}| = |\varepsilon_{\kappa\lambda}| = |\varepsilon| \qquad (B.13)$$

with $|\varepsilon| = 1$ from Eq.(A.12), the validity of Eq.(B.11) confirms that of Eq.(B.10).

No further proof of Eq.(B.9) is needed, in fact, for those transformations which can be considered to result from a continuous succession of infinitesimal canonical transformations, starting from the identity $X_\mu = \delta_{\mu\kappa}x_\kappa$ and, hence, from the value $J = 1$ of the jacobian.†

---

† And hence for the dynamical development of a hamiltonian system by Problem A.1.

It is not obvious, on the other hand, that the alternative value $J = -1$, admitted by Eq.(B.10), has to be excluded in the most general case of a canonical transformation. The sign of the jacobian determinant might be regarded as insignificant, since it reverses upon a mere interchange of two rows or columns. According to Eq. (B.6), however, this freedom does not exist if the volume-element of the phase space is to be understood as a positive quantity, irrespective of the variables chosen as coordinates.

The general proof of Eq.(B.8) can be obtained by considering first the determinant of an arbitrary matrix with elements $a_{\kappa\mu}(\kappa, \mu = 1, 2, \ldots, 2n)$. The determinant $|a|$ of this matrix is to be expanded into products of 2 by 2 subdeterminants

$$\begin{vmatrix} a_{2i-1,\mu_i} & a_{2i-1,\nu_i} \\ a_{2i,\mu_i} & a_{2i,\nu_i} \end{vmatrix} = a_{\kappa\mu_i}\varepsilon^i_{\kappa\lambda}a_{\lambda\nu_i} \qquad (B.14)$$

with $\varepsilon^i_{k\lambda}$ from Eq.(A.8) by summing over all permutations P for which the set of numbers

$$1, 2, \ldots, 2i - 1, 2i, \ldots, 2n - 1, 2n \qquad (B.15a)$$

goes over into the set

$$\mu_1, \nu_1, \ldots, \mu_i, \nu_i \ldots \mu_n, \nu_n \qquad (B.15b)$$

One thus obtains

$$|a| = \frac{1}{2^n} \sum_P c(P) \prod_{i=1}^n a_{\kappa\mu_i}\varepsilon^i_{\kappa\lambda}a_{\lambda\nu_i} \qquad (B.16)$$

where $c(P)$ is $+1$ for even and -1 for odd permutations. The factor $1/2^n$ accounts for the fact that, otherwise, each subdeterminant would be counted twice upon interchange of the columns μ_i and ν_i, resulting in a change of sign which is compensated by that of $c(P)$.

Upon replacement of $\varepsilon^i_{\kappa\lambda}$ and $\varepsilon_{\kappa\lambda}$ in the expression on the right side of Eq.(B.16) writing

$$\varepsilon_{\kappa\lambda} = \sum_{j_i=1}^n \varepsilon^{j_i}_{\kappa\lambda} \qquad (B.17)$$

according to Eq.(A.7) and interchanging the summation over the permutations and the summation over j_i, one finds

$$\frac{1}{2^n} \sum_P c(P) \prod_{i=1}^n a_{\kappa\mu_i}\varepsilon_{\kappa\lambda}a_{\lambda\nu_i} = \sum_{(j_i)} |a(j_i)| \qquad (B.18)$$

The sum on the right side extends here over all sets of numbers (j_i) obtained by letting each j_i independently assume any number from 1 to n, and the n^2 determinants $|a(j_i)|$, pertaining to these sets, are those of the matrices $a_{\kappa\lambda}(j_i)$ which arise from the matrix $a_{\kappa\lambda}$ upon replacement of $\kappa = 2i - 1$ and $\kappa = 2i$ by $\kappa = 2j_i - 1$ and $\kappa = 2j_i$ respectively. Consequently, all these determinants vanish for which the same number appears more than once in the set (j_i), and each of the remaining determinants, corresponding to a permutation of n *different* numbers j_i, is equal to the determinant $|a|$ of the original matrix $a_{\kappa\lambda}$.

It thus follows from Eq.(B.18), and equivalent to Eq.(B.16), that the determinant of an arbitrary even matrix $a_{\kappa\lambda}$ can be expressed in the form.

$$|a| = \frac{1}{2^n n!} \sum_P c(P) \prod_{i=1}^{n} a_{\kappa\mu_i} \varepsilon_{\kappa\lambda} a_{\lambda\nu_i} \qquad (B.19)$$

The jacobian J of a general transformation $X_\mu(x)$ with $a_{\kappa\mu} = \partial X_\mu / \partial x_\kappa$ can now be expressed as

$$J = -\frac{1}{2^n n!} \sum_P c(P) \prod_{i=1}^{n} [X_{\mu_i}, X_{\nu_i}] \qquad (B.20)$$

where the Poisson Bracket is defined by Eq.(A.23). For a canonical transformation one has therefore, in view of the condition of Eq.(A.28)

$$J = \frac{1}{2^n n!} \sum_P c(P) \prod_{i=1}^{n} \varepsilon_{\mu_i\nu_i} \qquad (B.21)$$

On the other hand, the application of Eq.(B.19) to the matrix $\varepsilon_{\kappa\lambda}$ yields for the determinant of that matrix

$$|\varepsilon| = \frac{1}{2^n n!} \sum_P c(P) \prod_{i=1}^{n} \varepsilon_{\kappa\mu_i} \varepsilon_{\kappa\lambda} \varepsilon_{\lambda\nu_i} \qquad (B.22)$$

From the properties of $\varepsilon_{\kappa\lambda}$ given in Eq.(A.11), however, it is seen that

$$\varepsilon_{\kappa\mu_i} \varepsilon_{\kappa\lambda} \varepsilon_{\lambda\nu_i} = \varepsilon_{\mu_i\nu_i} \qquad (B.23)$$

Substitution of this relation into Eq.(B.22) and comparision with Eq.(B.21) then leads to the result

$$J = |\varepsilon| \qquad (B.24)$$

By means of Eq.(A.12), one has therefore

$$J = 1 \qquad (B.25)$$

thus completing the proof of Eq.(B.9) for the jacobian of a general canonical transformation.

Appendix C

Molecular Distributions(*)

In dealing with the molecular constitution of material systems, it is indicated to regard the molecules as the subsystems considered in the analysis of Sec.5. For the present purposes they shall be treated as masspoints according to the discussion of Sec.4, so that the dynamical variables x_s of Appendix A stand here for the cartesian coordinates and the components of the momentum, assigned to individual molecules. A given set of these variables can be represented by the corresponding point in a 6-dimensional space, called the μ-space, and the choice of N such points, each pertaining to a specific molecule, is equivalent to that of a single representative point in the $6N$-dimensional phase space of the total system.

As a question of particular interest, one can ask for the number of molecules for which the corresponding points are contained in a given region of the μ-space. The quantity ϕ of Sec.5, chosen here to signify this number, has the additive property of Eq.(5.21) with the function ϕ_s of the dynamical variables (x_s) defined by

$$\phi_s = \begin{cases} 1 & \text{for } (x_s) \text{ within } r \\ 0 & \text{for } (x_s) \text{ outside of } r \end{cases} \tag{C.1}$$

Since in either case

$$\phi_s^2 = \phi_s \tag{C.2}$$

and therefore also

$$\overline{(\phi_s^2)} = \overline{\phi_s} \tag{C.3}$$

it then follows from Eq.(5.33) that

$$\overline{(\Delta \phi_s)^2} = \overline{\phi_s}(1 - \overline{\phi_s}) \tag{C.4}$$

It may be noted from the general expression of Eq.(5.4), for average values and in view of Eqs.(C.1) and (5.1), that

$$\overline{\phi_s} = P_R \tag{C.5}$$

where the region R in the phase space of the total system extends here over the region r in the phase space of the variables x_s and over the entire phase space of the variables $x_{s'}$ for $s' \neq s$. $\overline{\phi_s}$ therefore represents the probability, regardless of all other molecules, of finding the dynamical variables of the molecule s in the region r of the μ-space.

Without further specifications there is a wide latitude among solids, liquids, or gases for choosing the material system under consideration. The following discussion will be specifically directed towards the case of principal importance in the kinetic theory of gases where the density is sufficiently low for the interaction between molecules only to manifest itself in relatively short interruptions of their free motion during a collision. The system is thus considered, in effect, to be an ideal gas with an energy given by the total kinetic energy of all its molecules.

In line with the simplifying procedure of the preceding paragraph, it will be assumed that the molecules are statistically uncorrelated and not distinguished from each other. Applying the ensuing results, it follows from Eq.(C.4), with the subscript s omitted and in the notation of Eqs.(5.35) and (5.39), that

$$(\Delta\varphi)\text{rms} = \sqrt{\bar{\varphi}(1 - \bar{\varphi})} \qquad (C.6)$$

Upon further insertion of

$$\bar{\varphi} = \bar{\phi}/N \qquad (C.7)$$

from Eq.(5.37) on the right side of Eq.(5.40), one thus obtains the relation

$$\frac{(\Delta\phi)\text{rms}}{\bar{\phi}} = \sqrt{\frac{1 - \bar{\phi}/N}{\bar{\phi}}} \qquad (C.8)$$

for the quantity ϕ, defined as the number of molecules with dynamical variables within the region r of the μ-space.

As a test, this region can be chosen to comprise the entire μ-space. According to Eq.(C.1) one has then $\phi_s = 1$ throughout, therefore from Eq.(5.21) $\phi = N$ and, hence, also $\bar{\phi} = N$. The vanishing result for $(\Delta\phi)\text{rms}$, obtained in this case from Eq.(C.8), signifies according to the discussion of Sec. 5, that any deviation of ϕ from its mean value is with certainty ruled out in agreement with the fact that ϕ is here identical with the fixed total number N of molecules.

It is of greater interest, however, to consider the opposite case, where $\bar{\phi} \ll N$ so that Eq.(C.8) can be replaced by

$$\frac{(\Delta\phi)\text{rms}}{\bar{\phi}} \simeq \frac{1}{\sqrt{\bar{\phi}}} \qquad (C.9)$$

In order to still assign to ϕ the essentially unique value $\bar{\phi}$, on the other hand, the criterion of Eq.(5.20) requires that we have $\bar{\phi} \gg 1$. Combined, one therefore deals with the condition

$$\bar{\phi}/N \ll 1 \ll \bar{\phi} \qquad (C.10)$$

which can be satisfied provided that N is sufficiently large. This case is particularly relevant when the region r is chosen so small that it can in effect be treated as a volume element $d\mu$ of the μ-space. Using the notation

$$\bar{\phi} = dN \qquad (C.11)$$

for the mean number of molecules within $d\mu$, let

$$dN = fd\mu \qquad (C.12)$$

where f is a function of the variables which designate the location of the volume element $d\mu$ in the μ-space. One has therefore

$$\int fd\mu = N \qquad (C.13)$$

where the integral extends over the entire μ-space.

To the extent to which the number of molecules within $d\mu$ can be expected with high confidence to deviate by only negligible amount from the mean number dN, the function f in effect uniquely determines the actual distribution of molecules in the μ-space. With this understanding, however, dN and $d\mu$ in Eq.(C.12) are not true differentials in the sense that they represent arbitrarily small quantitites. While $d\mu$ can be so small for a given finite value of f to have $dN << N$, it must be chosen large enough to yield $dN >> 1$ in order to satisfy the condition of Eq.(C.10) in the notation of Eq.(C.11). This is possible for sufficiently large values not only of N but also of f.

With the understanding of the preceding discussion, the distribution function contains all the information required to deduce macroscopic properties and is therefore the primary object of investigation in the kinetic theory of gases. A key to the investigation is furnished by the equation for f due to Boltzmann, which formulates the rate of change under the effect of collisions between the molecules. In particular, this equation is found to have a stationary solution, characterized by an exponential dependence on the kinetic energy of a molecule in accordance with the Maxwell distribution of velocities for a gas in thermal equilibrium. One deals here with a special case of considerations, later introduced to arrive at the general condition of thermal equilibrium. This will be achieved by first considering the time-dependence of the distribution in the phase space (Section 6) and then proceeding to establish the requirements which it must satisfy in order to expect stationary properties of macroscopic systems (Section 7).

Appendix D

Some Properties of Fourier Series

Several useful properties of finite Fourier series follow from the finite geometric series

$$S_N(x) = \sum_{s=1}^{N} x^s = x \left[\frac{x^N - 1}{x - 1} \right] \tag{D.1}$$

which implies

$$S_N(\theta) = \sum_{s=1}^{N} e^{i\theta s} = e^{i\theta} \left[\frac{e^{iN\theta} - 1}{e^{i\theta} - 1} \right] \tag{D.2}$$

For example, if $\theta = 2\pi(m - n)/N$, where m and n are integers with $|m| < N/2$ and $|n| < N/2$, then

$$\frac{1}{N} \sum_{s=1}^{n} \exp\left(\frac{2\pi i m s}{N} \right) \exp\left(\frac{-2\pi i n s}{N} \right) = \delta_{mn} \tag{D.3}$$

The result for $m = n$ follows by explicit summation. Expressions for *sine* and *cosine* sums can be obtained by writing

$$\cos\theta = \frac{e^{i\theta} + e^{-i\theta}}{2} \quad ; \quad \sin\theta = \frac{e^{i\theta} - e^{-i\theta}}{2i} \tag{D.4}$$

and using the above. For example

$$\sum_{s=1}^{N} \cos\theta_m s \cos\theta_n s = \frac{1}{4} \sum_{s=1}^{N} \left[e^{i(\theta_m + \theta_n)s} + e^{-i(\theta_m + \theta_n)s} + e^{i(\theta_m - \theta_n)s} \right.$$

$$\left. + e^{-i(\theta_m - \theta_n)s} \right] \tag{D.5}$$

It follows that if $\theta_m = 2\pi m/N$ and $\theta_n = 2\pi n/N$ where m and n are integers, and if $|m| < N/2$ and $|n| < N/2$, then

$$\frac{2}{N} \sum_{s=1}^{N} \cos\frac{2\pi m s}{N} \cos\frac{2\pi n s}{N} = \delta_{mn} \tag{D.6}$$

The result for $m = n$ follows by explicitly summing the last two terms in Eq.(D.5). In a similar fashion one shows that

$$\frac{2}{N} \sum_{s=1}^{N} \sin\frac{2\pi m s}{N} \sin\frac{2\pi n s}{N} = \delta_{mn} \tag{D.7}$$

$$\frac{2}{N} \sum_{s=1}^{N} \sin\frac{2\pi m s}{N} \cos\frac{2\pi n s}{N} = 0 \tag{D.8}$$

under these conditions.

Equation (D.2) can be rewritten in the form

$$S_N(\theta) = \sum_{s=1}^{N} e^{i\theta s} = \frac{e^{i(N+1)\theta} - 1}{e^{i\theta} - 1} - 1 \qquad (D.9)$$

Now consider the *sine* sum

$$\sum_{s=1}^{N} \sin\theta_m s \sin\theta_n s = -\frac{1}{4} \sum_{s=1}^{N} \left[e^{i(\theta_m+\theta_n)s} + e^{-i(\theta_m+\theta_n)s} - e^{i(\theta_m-\theta_n)s} \right.$$

$$\left. - e^{-i(\theta_m-\theta_n)s} \right]$$

$$(D.10)$$

and let $\theta_m = \pi m/(N+1)$ and $\theta_n = \pi n/(N+1)$ where m and n are positive integers satisfying the conditions $m < N+1$ and $n < N+1$. It follows that

$$\frac{2}{N+1} \sum_{s=1}^{N} \sin\frac{\pi m s}{N+1} \sin\frac{\pi n s}{N+1} = \delta_{mn} \qquad (D.11)$$

The term with $m = n$ follows by using Eq.(D.9) on the first two terms in Eq.(D.10) and explicitly summing the last two.

Appendix E
Basic Texts and Monographs

Binder, K., ed., *Monte Carlo Methods in Statistical Physics*, 2nd ed., Springer-Verlag, New York (1986)

Creutz, M., *Quarks, Gluons, and Lattices*, Cambridge University Press, Cambridge (1983)

Davidson, N., *Statistical Mechanics*, McGraw-Hill Book Company, New York (1965)

Domb, G., and J. L. Lebowitz, eds., *Phase Transitions and Critical Phenomena*, vol. 9, Academic Press, New York (1984); see also previous volumes.

Fetter, A. L., and J. D. Walecka, *Quantum Theory of Many-Particle Systems*, McGraw-Hill Book Company, New York (1971)

Gibbs, J. W., *Elementary Principles in Statistical Mechanics*, Yale University Press (1902); reprinted by Dover Publications, New York (1960)

Fowler, R. H., and E. A. Guggenheim, *Statistical Thermodynamics*, rev. ed., Cambridge University Press, Cambridge (1949)

Hill, T. L., *An Introduction to Statistical Thermodynamics*, Addison Wesley Publishing Company, Reading, Mass. (1960)

Huang, K., *Statistical Mechanics*, John Wiley and Sons, Inc., New York (1963)

Kittel, C., *Elementary Statistical Physics*, John Wiley and Sons, Inc., New York (1958)

Kubo, R., *Statistical Mechanics, An Advanced Course with Problems and Solutions*, Interscience, New York (1965)

Landau, L. D., and E. M. Lifshitz, *Statistical Physics*, 2nd ed., Addison-Wesley Publishing Company, Reading, Mass. (1969); 3rd ed., Pergamon Press, London (1980)

Lichtenberg, A. J., and M. A. Lieberman, *Regular and Stochastic Motion*, Springer-Verlag, New York (1983)

Ma, S. K., *Statistical Mechanics*, World Scientific, Singapore (1985)

Mayer, J. E., and M. G. Mayer, *Statistical Mechanics*, 2nd ed., John Wiley and Sons, Inc., New York (1977)

Negele, J. W., and H. Orland, *Quantum Many-Particle Systems*, Addison-Wesley Publishing Company, Reading, Mass. (1988)

Reif, F., *Fundamentals of Statistical and Thermal Physics*, McGraw-Hill Book Company, New York (1965)

Rushbrook, G. S., *Introduction to Statistical Mechanics*, Oxford University Press, Oxford (1949)

Schrödinger, E., *Statistical Thermodynamics*, 2nd ed., Cambridge University Press, Cambridge (1957)

Stanley, H. E., *Introduction to Phase Transitions and Critical Phenomena*, Clarendon Press, Oxford (1971)

Ter Haar, D., *Elements of Thermostatistics*, 2nd ed., Holt, Reinhart and Winston, New York (1966)

Tolman, R. C., *The Principles of Statistical Mechanics*, Oxford University Press, Oxford (1938); reprinted by Dover Publications, Inc., New York (1979)

Wannier, G. H., *Statistical Physics*, John Wiley and Sons, New York (1966)

Zaslavsky, G. M., *Chaos in Dynamic Systems*, Harwood Academic Publishers, New York (1985)

Problems

Problems

1.1 Consider the coordinate transformation $x = r \cos \phi, y = r \sin \phi$, and find the corresponding transformation $p_{x,y} = p_{x,y}(p_r, p_\phi, r, \phi)$, where p_r and p_ϕ are the momenta conjugate to r and ϕ respectively. Show that p_ϕ represents the angular momentum $x p_y - y p_x$, determine the jacobian of the transformation $(p_x, p_y, x, y) \longrightarrow (p_r, p_\phi, r, \phi)$, and express $p_x^2 + p_y^2$ in terms of p_r, p_ϕ, r, ϕ.

1.2 Use the transformation

$$x = r \sin \theta \cos \phi; \quad y = r \sin \theta \sin \phi; \quad z = r \cos \theta$$

from the set (x, y, z) of cartesian coordinates to the set (r, θ, ϕ) of polar coordinates to express the momenta p_x, p_y, p_z, conjugate to x, y, z, as functions of the polar coordinates r, θ, ϕ and their conjugate momenta p_r, p_θ, p_ϕ. Then proceed to determine as functions of the same variables the following quantities: $p^2 = p_x^2 + p_y^2 + p_z^2$, the components a_x, a_y, a_z of the angular momentum vector $\mathbf{a} = [\mathbf{r} \wedge \mathbf{p}]$, and $a^2 = a_x^2 + a_y^2 + a_z^2$.

1.3 By means of the coordinate transformation

$$x = r \sin \theta \cos \phi; \quad y = r \sin \theta \sin \phi; \quad z = r \cos \theta$$

express the hamiltonian $H = (p_x^2 + p_y^2 + p_z^2)/2m + U(x, y, z)$ in terms of r, θ, ϕ and their conjugate momenta p_r, p_θ, p_ϕ. Assuming that U depends only on r, show from Hamilton's equations that one may assume $p_\theta = 0, \theta = \pi/2$ and that the hamiltonian has then the form obtained for polar coordinates in the plane $z = 0$.

1.4 Show that the volume element in the phase space is invariant against a coordinate transformation $q_k' = q_k'(q_i)$ and the corresponding canonical transformation $p_k' = p_k'(p_i, q_i)$ of the momenta in Eq.(1.49) in the sense that $d\lambda = d\lambda'$, where $d\lambda = \prod_i dp_i dq_i$ and $d\lambda' = \prod_k dp_k' dq_k'$.

1.5 The hamiltonian for two particles with masses m_1 and m_2 moving in a plane shall be given by

$$H = (p_{x_1}^2 + p_{y_1}^2)/2m_1 + (p_{x_2}^2 + p_{y_2}^2)/2m_2 + V(r)$$

where $V(r)$, the potential energy, shall only depend upon the relative distance

$$r = [(x_1 - x_2)^2 + (y_1 - y_2)^2]^{1/2}$$

and where the four cartesian coordinates x_1, y_1, x_2, y_2, are used as coordinates.

(a) Using the new coordinates,

$$X = \frac{m_1 x_1 + m_2 x_2}{m_1 + m_2} \qquad\qquad y = \frac{m_1 y_1 + m_2 y_2}{m_1 + m_2}$$

$$r = [(x_1 - x_2)^2 + (y_1 - y_2)^2]^{1/2} \qquad \phi = \text{arctg}\, \frac{y_1 - y_2}{x_1 - x_2}$$

find the corresponding momenta p_X, p_Y, p_r, p_ϕ.

(b) show that p_X and p_Y are the components of the total momentum and that p_ϕ is the angular momentum around the center of gravity.

(c) Find the hamiltonian $H' = H'(X, Y, r, \phi, p_X, p_Y, p_r, p_\phi)$, derive Hamilton's equations in the new variables, and show that p_X, p_Y, and p_ϕ are constants of the motion.

3.1 Let $H = p^2/2m + m\omega^2 q^2/2$ (harmonic oscillator).

(a) Determine the values of p and q at a time t in terms of their values p_0 and q_0 at the time $t = 0$ and the derivatives

$$\frac{\partial p}{\partial p_0}; \ \frac{\partial p}{\partial q_0}; \ \frac{\partial q}{\partial p_0}; \ \frac{\partial q}{\partial q_0}$$

(b) Use the preceding results to show explicitly that the jacobian has the value $J = 1$ and to demonstrate the conservation of phase volume (Liouville theorem) with the help of geometric arguments.

3.2 Two functions $y(x)$ and $z(x)$ satisfy the following set of differential equations and initial conditions

$$\frac{dy}{dx} = f(y, z) \qquad y(0) = y_0$$

$$\frac{dz}{dx} = g(y, z) \qquad z(0) = z_0$$

Consider the jacobian for a transformation from the variables y, z to y_0, z_0.

$$\Delta = \begin{vmatrix} \dfrac{\partial y}{\partial y_0} & \dfrac{\partial y}{\partial z_0} \\ \dfrac{\partial z}{\partial y_0} & \dfrac{\partial z}{\partial z_0} \end{vmatrix}$$

Find the necessary and sufficient conditions for $d\Delta/dx = 0$.

5.1 Derive $P_R \cong \exp(-C^2/2)/C$, where P_R and C are defined by Eqs.(5.1) and (5.13) respectively, if there is a gaussian dependence of finding ϕ within an interval $d\phi$.

5.2 From the probability

$$P_{N_r} = \binom{N}{N_r} (P_r)^{N_r} (1 - P_r)^{N-N_r}$$

of finding N_r out of N molecules within a region r of the μ-space, calculate:

(a)

$$\bar{N}_r = \sum_{N_r=0}^{N} N_r P_{N_r}$$

(b)

$$\overline{(N_r^2)} = \sum_{N_r=0}^{N} N_r^2 P_{N_r}$$

and hence

(c)

$$(\Delta N_r)_{rms}/\bar{N}_r$$

with

$$(\Delta N_r)_{rms} = \left[\overline{(N_r - \bar{N}_r)^2} \right]^{1/2}$$

(**Hint:** Consider the function $f(x) = \sum_{N_r=0}^{N} x^{N_r} P_{N_r}$ and its derivatives.) You will find the material in Appendix C useful in working this problem.

6.1 The probability density ρ for a particle moving in the x-direction with momentum p and hamiltonian $H = p^2/2m$ shall at the time t_0 be given by

$$\rho(p, x, t_0) = C \exp \left\{ - \left[\alpha(p - p_0)^2 + \beta(x - x_0)^2 \right] \right\}$$

where $\alpha > 0$ and $\beta > 0$. For an arbitrary time t, determine the corresponding expression $\rho(p, x, t)$ for ρ, the mean values $\bar{x}$ and $\bar{p}$, as well as the mean squares of $x - \bar{x}$ and $p - \bar{p}$.

6.2 Show that an arbitrary coordinate transformation $Q_k = Q_k(q)$ or $q_i = q_i(Q)$ from the set q_i to the set Q_k $(i, k = 1, 2, \dots, n)$ of coordinates satisfies the condition of a canonical transformation from the set (p, q) to the set (P, Q) stated in Eqs.(6.27) if the momenta P_k, conjugate to

Q_k, are obtained from the momenta p_i, conjugate to q_i, by means of the relations

$$P_k = \sum_{i=1}^{n} p_i \frac{\partial q_i}{\partial Q_k}$$

6.3 Show for an arbitary hamiltonian $H(p_s, q_s)$ that Hamilton's equations

$$\frac{dp_s}{dt} = -\frac{\partial H}{\partial q_s} \quad ; \; s = 1, 2, \ldots, f$$
$$\frac{dq_s}{dt} = \frac{\partial H}{\partial p_s}$$

remain valid if both sides of these equations are replaced by their ensemble averages with a distribution function $D(p_s, q_s)$ which vanishes for $|p_s| \longrightarrow \infty$ and $|q_s| \longrightarrow \infty$.

7.1 Given the average values of p, x, p^2, x^2, and xp at the time $t = 0$ for an ensemble of mass points with mass m, freely moving in the x-direction, use Eq.(7.14) to find the corresponding averages and the mean square deviation of p and x from their average at the time t, and verify that the latter are positive.

7.2 Assume that the distribution function for a free particle in one-dimensional motion with hamiltonian $H = p^2/2m$ has at the time t_0 the form

$$D = \frac{C}{1 + \alpha p^2 + \beta q^2}$$

where α and β shall both be positive. Following each point in the phase space as a function of time, find the distribution at a later time t. Find the conditions under which D is stationary and show that they are equivalent to the conditions necessary, in order that D has the same value along a phase orbit.

7.3 Assume that the stationary distribution function for the one-dimensional harmonic oscillator has the form

$$D = C \exp\left[-\beta \left(\frac{p^2}{2m} + \frac{m}{2}\omega^2 q^2\right)\right]$$

where β is a positive constant. Find the probability density $\rho = \rho(p, q)$ and the expectation values for p, q, p^2, q^2, and $H = p^2/2m + m\omega^2 q^2/2$.

8.1 Consider a system composed of two one-dimensional harmonic oscillators with frequencies $\alpha\omega_0$ and $\beta\omega_0$ (α and β are positive integers different from zero).

$$H = H_1 + H_2$$

$$H_1 = \frac{1}{2m}p_1^2 + \frac{1}{2}m\alpha^2\omega_0^2 q_1^2$$

$$H_2 = \frac{1}{2m}p_2^2 + \frac{1}{2}m\beta^2\omega_0^2 q_2^2$$

(a) Two of the constants of the motion may be considered to be $c_1 = H_1$ and $c_2 = H_2$. Show that $c_3 = Im(\pi_1)^\beta(\pi_2^*)^\alpha$ is a third constant of the motion, where

$$\pi_1 = \frac{1}{\sqrt{m\omega_0}}p_1 + i\alpha\sqrt{m\omega_0}q_1$$

$$\pi_2^* = \frac{1}{\sqrt{m\omega_0}}p_2 - i\beta\sqrt{m\omega_0}q_2$$

and verify that in the special case $\beta = \alpha$ it reduces to $c_3 = p_2 q_1 - p_1 q_2$.

(b) Let c_1, c_2, and q_2 have certain fixed values and let $\beta = \alpha + 1$ be an even number. How many times does c_3 go through its full range of values as q_1 varies *once* through its full range of values?

8.2 A special solution of the two-dimensional harmonic oscillator shall have the form

$$x = a_1 \sin\omega_1 t \qquad\qquad y = a_2 \sin\omega_2 t$$

$$p_x = m\omega_1 a_1 \cos\omega_1 t \qquad p_y = m\omega_2 a_2 \cos\omega_2 t$$

Let $\omega_2 = \omega_1(Z+1)/Z$, where Z is a positive integer.

(a) Give the number of points at which the orbit in the x-y plane intercepts the y-axis and the largest distance δ which occurs between two consecutive points as a function of Z

(b) How large does Z have to be chosen until $\delta < a_2/100$, and how long will it take, in this case, until each of the intercepting points with the y-axis has been reached at least once?

(c) For Z sufficiently large, does the system become quasiergodic? Justify your answer.

10.1 Use the analysis of Section 1 to show that the relations

$$p = \alpha\sqrt{2p'}\cos q' \quad ; \quad q = \frac{1}{\alpha}\sqrt{2p'}\sin q'$$

represent a canonical transformation $(p, q) \longleftrightarrow (p', q')$. Expressing the hamiltonian $H = p^2/2m + m\omega^2 q^2/2$ as a function of p' and q', show that it can be made independent of q' by suitable choice of α, and use this form of the hamiltonian to determine its mean value E at the temperature T according to classical statistics.

12.1 By means of the partition function $Z = \int \exp(-\beta H) \, d\lambda$ of a system with the arbitary hamiltonian H and with $E = \bar{H}, (\Delta E)_{\text{rms}} = \left[\overline{(H - \bar{H})^2}\right]^{1/2}$, determine the general form of the function $E = E(\beta)$ such that $(\Delta E)_{\text{rms}}/E$ has a value c, independent of β.

12.2 Given a hamiltonian $H = T + U$

$$T = \sum_{r=1}^{f} \frac{1}{2m} p_r^2$$

$$U = U(q_1, q_2 \ldots, q_f)$$

for a system in thermal equilibrium with f degrees of freedom.

   (a) Find $\pi(T_1)$ such that $\pi(T_1)dT_1$ = probability of finding T between T_1 and $T_1 + dT_1$

   (b) Find $\bar{T}$

   (c) Show that for $f \gg 1$, one has $\pi(T) = c_1 \exp\left[-c_2(T - \bar{T})^2\right]$. Find the values of c_1 and c_2.

12.3 The hamiltonian of a system of N particles is $H = K + U$, where

$$K = \sum_{i=1}^{f} \frac{1}{2m_i} p_i^2$$

represents the kinetic energy ($f = 3N$) and where the potential energy $U = U(q_1, q_2, \ldots, q_f)$ shall be an arbitrary function of the coordinates. With the system in equilibrium at the temperature T and denoting by $P(K)dK$ the probability that the value of the kinetic energy lies between K and $K + dK$, determine the function $P(K)$ (see Prob. 12.2) and use it to determine $\bar{K}$ and

$$(\Delta K)_{\text{rms}}/\bar{K} = \left[\overline{(\Delta K)^2}\right]^{1/2}/\bar{K}$$

13.1 Derive from the canonical distribution for an ideal gas of N molecules contained in a volume V the average number $\bar{n}$ of molecules contained in a volume $v < V$ and the root mean square of deviation

$$\Delta n = \left[\overline{(n - \bar{n})^2}\right]^{1/2}$$

assuming both $N \gg 1$ and $n \gg 1$.

13.2 A system of N one-dimensional uncoupled harmonic oscillators of mass m and circular frequency ω shall be under the influence of a constant external force F, acting on each of them. Find the amount of work ΔW required in an isothermal reversible process to change F from zero to F^* and the amount of heat ΔQ gained by the system in this process. Show that the result agrees with the change of entropy in that process.

13.3 N molecules constitute an ideal gas of volume V in equilibrium. What is the probability P_n of finding a given number n of them within a volume $\tau < V$? Apply the result to obtain $\bar{n}$ and $(\Delta n)_{\text{rms}}$ (**Hint:** Use the binomial theorem) and determine $(\Delta n)_{\text{rms}}/\bar{n}$ for a mole volume of 22.4 liters, assuming $\tau = 1\text{cm}^3$ so that $\tau \ll V$. (Compare Prob.5.2.)

13.4 For an ideal gas of particles with rest mass m_0 and kinetic energy $c[p^2 + (m_0c)^2]^{1/2} - m_0c^2$, determine the energy ε and the specific heat c_v per mole

(a) For $\varepsilon = k_BT/m_0c^2 \ll 1$, including terms linear in ε

(b) For $\gamma = m_0c^2/k_BT \ll 1$, including linear and quadratic terms in γ

13.5 Show for an arbitrary system that the mean square deviation of the energy from its average value E is given by

$$\overline{(\Delta E)^2} = -\frac{\partial E}{\partial \beta}$$

and apply the result to find $(\Delta E)_{\text{rms}}/E$ for an ideal gas of N atoms.

13.6 An ideal diatomic gas of absolute temperature T, consisting of N molecules, shall be contained in a cylindrical vessel with its axis in the z direction and a solid base of area τ at $z = 0$. The molecules shall all have $z \geq 0$, and they shall be under the influence of a gravitational force mg ($m = $ mass of molecule), acting in the negative z-direction. Find the expectation value E of the energy of the gas and the specific heat $C = \partial E/\partial T$.

13.7 For the gas described in Prob.13.6 find the expectation value of the density $\rho(z)$ as a function of z and particularly the ratio $r = \rho(z)/\rho(0)$ in terms of the molecular weight M of the gas and gas constant R. Find the value $z_{1/2}$ of z for which $r = 1/2$, assuming $T = 273$ centidegrees and assuming the gas to consist

(a) of oxygen (O_2)

(b) of hydrogen (H_2)

(Take $g = 981$ cm/sec^2 and $R = 8.31 \times 10^7$ ergs/degree.)

13.8 What is the density and specific heat of a mole of an ideal gas at temperature T contained in a volume V if each molecule is subject to the same constant force F in the x-direction?

13.9 Show that the mean value $\bar{\rho}$ of the density of an ideal gas under the action of a potential $U = U(x, y, z)$ in a canonical ensemble is given by $\bar{\rho} = \rho_0 \exp(-U/k_B T)$. Give the deviation h_y for which the density of a gas with molecular weight M above the earth's surface has reached $1/e^{th}$ of its value at sea level for $M = 44$, $M = 32$, $M = 28$, and $M = 2$.

13.10 Suppose a very big number N of molecules to be dropped statistically into a box which is divided into two cells 1 and 2 of the same size. Let n_1 and n_2 be the number of molecules in cell 1 and cell 2 respectively $(n_1 + n_2 = N)$.

(a) Give the rigorous expression of the probability $P(n_1)$ of finding a certain number n_1 in cell 1

(b) Show that this probability has a maximum for $n_1 = N/2$

(c) Give, with the use of Stirling's formula†, an asymptotic expression of $P(n)$ valid for very big numbers N, n_1, and n_2

(d) Calculate with the help of the asymptotic formula the probability $P(\delta)$ of finding $n_1 = N/2 + \delta, n_2 = N/2 - \delta$, supposing $N \gg \delta \gg 1$

(e) Let $N = 10^{13}$ (order of magnitude of the number of air molecules in 1 cm^3). Calculate the numerical value (using the asymptotic expression) of $P(\delta)$ for

$$i) \qquad \delta = 10^6$$
$$ii) \qquad \delta = 10^{10}$$
$$iii) \qquad \delta = 10^{14}$$

13.11 A gas of $N \gg 1$ molecules with mass m stands under the action of a potential energy $U(\mathbf{r})$ per molecule. N_i shall be the number of molecules with a position vector $\mathbf{r}_i$ within the range $d\mathbf{r}$ and a velocity vector $\mathbf{v}_i$ within the range $d\mathbf{v}$. Calculate those values of N_i for which the distribution number has a maximum with the restrictions

$$\sum_i N_i = N = \text{const.}$$

$$\sum_i N_i \left[\frac{1}{2} m v_i^2 + U(\mathbf{r}_i) \right] = E = \text{const.}$$

---

† Stirling's formula is $lnN! \sim (N+1/2)lnN - N + (1/2)ln(2\pi) + O(1/N)$

and use Stirling's formula†, supposing the values of N_i still to be very big numbers, and show that these values are given by

$$N_i^* = A \exp\left\{-\beta\left[\frac{1}{2}mv_i^2 + U(\mathbf{r}_i)\right]\right\}$$

A and β being constants.

14.1 Consider an ideal monatomic gas of A (Avogadro's number) molecules enclosed in a cylindrical tube of radius ρ with the axis parallel to the x direction and closed with two end walls at $x = 0$ and $x = \ell$. The walls shall be represented by a very sharply rising potential energy.

(a) Show that

$$\frac{1}{\ell}\sum_{i=1}^{A}\left(x_i\frac{\partial H}{\partial x_i}\right) = \bar{F}$$

represents the average value of the force exerted by the molecules on the wall at $x = \ell$

(b) Derive from the fact proven in (a) that the pressure on this wall is given by

$$p = \frac{RT}{V}$$

where V is the volume of the container and R is the gas constant.

15.1 The motion of a mass point moving in a plane shall be described by the hamiltonian

$$H = \frac{1}{2m}p_x^2 + \frac{1}{2m}p_y^2 + U(x, y)$$

where $U(x, y) = A(x^2 - xy + 3y^2)^2$ with $A > 0$.

(a) Show by direct substitution that

$$x\frac{\partial U}{\partial x} + y\frac{\partial U}{\partial y} = 4U$$

(b) Use the fact proven in (a) to show that $U = k_BT/2$

16a.1 Given the partition function $Z(\beta)$ for an arbitrary system, find the mean square deviation $\overline{(\Delta E)^2}$ for the energy from its average E. With $\beta = 1/k_BT$, determine the general form of the function $E(T)$ such that

$$\sqrt{\overline{(\Delta E)^2}}/E$$

is independent of T, and show that it is consistent with the properties of a monatomic ideal gas. (Compare Prob.12.1.)

16a.2 N particles with negligible interaction shall be confined within a volume V. The potential energy of a particle shall have the value $U = 0$ in a part $V' < V$ of the volume and the value $U = a$ in the remaining part $V - V'$. With the system in equilibrium at a constant temperature T, find the increase ΔE of its energy and the amount of work ΔW required in a change from $a = 0$ to $a = a^*$. Compare the results for $a^* = \infty$ with those obtained in the isothermal compression of an ideal gas.

16b.1 The potential energy of a diatomic molecule shall be given by $U(r) = \kappa(r - r_0)^2$, where r is the distance between the two atoms. Neglecting only quadratic and higher terms in $1/\kappa$, find the specific heat c_v per mole for an ideal gas of these molecules and the mean value $\bar{r}$ of the distance r at the absolute temperature T.

16b.2 The specific heat per mole of an ideal diatomic gas at the temperature T can be represented by the power series $c_v(T) = c_0 + c_1 T + c_2 T^2 + \dots$. On the basis of classical statistics, determine the coefficients c_0 and c_1 in terms of the expansion coefficients of $U(r^* + \rho)$ in powers of ρ, where $U(r)$ is the potential energy of the molecule, r the distance between the atoms, and r^* the equilibrium value of r. Assuming that U can be considered as a quadratic function of ρ, comment upon the result in regard to the applicability of the equipartition theorem.

16b.3 The total energy of a diatomic molecule shall be given by

$$T + U = \frac{1}{2}m_1(\dot{x}_1^2 + \dot{y}_1^2 + \dot{z}_1^2) + \frac{1}{2}m_2(\dot{x}_2^2 + \dot{y}_2^2 + \dot{z}_2^2) + U(r)$$

where m_1, x_1, y_1, z_1 and m_2, x_2, y_2, z_2 are the masses and coordinates of atoms 1 and 2 respectively and where r is the relative distance between them.

(a) Set up the hamiltonian H_{mol} for the molecule using the coordinates of the center of gravity X, Y, Z, the relative distance r, and the angles θ and ϕ.

(b) Assuming the potential energy $U(r)$ to have the form

$$U = A\left[\left(\frac{r_0}{r}\right)^8 - \left(\frac{r_0}{r}\right)^6\right] \quad ; A > 0$$

find the value r^* for which U has its minimum and give an expression of U of the form

$$U = U_{\text{min}} + B(r - r^*)^2 + C(r - r^*)^3$$

where the quantities $U_{\min}$, B, and C shall be expressed in terms of A and r_0.

(c) Neglecting the cubic and higher terms in U and the variation in the moment of intertia due to deviations of r from r^*, and assuming $A \gg k_B T$, calculate the root mean square

$$\sqrt{(r - r^*)^2} \equiv \Delta$$

and the specific heat of N molecules at the absolute temperature T

(d) As an estimate of the importance of the cubic term, calculate $C\Delta^3$ and determine a temperature T^* such that for $T < T^*$ one finds

$$C\Delta^3 < \frac{1}{10} B\Delta^2$$

16b.4 Calculate the specific heat of a diatomic and a triatomic gas, allowing for a vibratory motion of the atoms within the molecule.

16b.5 Prove Eq.(16.6b).

16b.6 Prove that the transformation in Eqs.(16.1b) and (16.2b) is canonical.

16b.7 Prove Eq.(16.8b) explicitly using the transformation to spherical polar coordinates derived in Sec. 1.

16c.1 Classically, the specific heat of a system composed of N one-dimensional, non-interacting harmonic oscillators is $C_V = N k_B$. Calculate classically, to lowest order in ϵ, the change in C_V due to the addition of an anharmonic term to the hamiltonian

$$H = \frac{1}{2m} p^2 + \frac{1}{2} m\omega^2 q^2 + \epsilon q^4$$

State explicitly under what conditions your expansion is valid.

16c.2 Let

$$H = \frac{1}{2m}(p_1^2 + p_2^2) + \frac{1}{2} m\omega^2(q_1^2 + q_2^2)$$

and let

$$p_1' = \frac{1}{\sqrt{2}}(p_1 + p_2) \qquad q_1' = \frac{1}{\sqrt{2}}(q_1 + q_2)$$

$$p_2' = \frac{m\omega}{\sqrt{2}}(q_1 - q_2) \qquad q_2' = -\frac{1}{m\omega\sqrt{2}}(p_1 - p_2)$$

Show that the hamiltonian H as well as Hamilton's equations have the same form if $p_1, q_1;\ p_2, q_2$ are replaced by p_1', q_1', p_2', q_2' respectively.

16c.3 Use the Schmidt orthogonalization procedure to demonstrate that one can still impose the condition of Eq.(16.32c) even if two of the eigenvalues are degenerate.

16c.4 Show that all other choices of l in Eq.(16.83c) reproduce one of the normal-mode solutions which has already been obtained.

16c.5 Show that all other solutions to the periodicity condition of Eq.(16.119c) reproduce one of the solutions already obtained through Eq.(16.120c).

16d.1 Compute the entropy of black-body radiation contained in a cubical box with volume Ω. What is the increase of the temperature T of the radiation if the volume in an adiabatic process is compressed to half its value?

16d.2 Demonstrate that Eq.(16.17d) leads to Eq(16.11d) when the fields in Eqs.(16.6d) are obtained from the normal-mode expansion in Eq.(16.12d).

17.1 Calculate the change of temperature dT of an ideal gas of N molecules with magnetic moments μ, if the external magnetic field is changed in an adiabatic process from $\mathcal{H}$ to $\mathcal{H} + d\mathcal{H}$. Express the result in terms of $\mathcal{H}, \mu, T$, and the specific heat C_V^0 which the gas would have for $\mathcal{H} = 0$.

17.2 A diatomic molecule shall be assumed to be rigid and to have a magnetic moment $\boldsymbol{\mu}$ of fixed magnitude μ parallel to the line of connection between the two atoms, causing its energy in the presence of a magnetic field $\mathbf{H}$ to change by the amount $-\boldsymbol{\mu} \cdot \mathbf{H}$. On the basis of the classical expression for the partition function, find the magnetic moment M, the energy E, and the specific heat at constant volume C_V for an ideal gas of N such molecules at the temperature T. With $|\mathbf{H}| \cong 10^4, \mu \cong 10^{-20}$ (both in c.g.s. units) and $T \cong 300° K$, estimate the relative correction to C_V due to the magnetic field.

17.3 The hamiltonian of a spin system of N atoms with magnetic moment μ and angular momentum $\hbar I$ shall be of the form

$$H = H_0 + D$$

where

$$H_0 = -\sum_{j=1}^{N} \boldsymbol{\mu}_j \cdot \mathbf{H}$$

represents the part due to a homogeneous magnetic field $\mathbf{H}$ of magnitude $\mathcal{H}$ and where

$$D = \frac{1}{2} \sum_{j \neq k} \sum_{r,s=1}^{3} T_{rs}^{jk} \mu_j^r \mu_k^s$$

represents the dipolar interaction between different atoms characterized by given tensor components T_{rs}^{jk} and

$$\mu_j^{1,2,3} = (\boldsymbol{\mu}_j)_{x,y,z}$$

(a) Find an expression for the partition function

$$\mathcal{Z} = Tr(-\beta H)$$

valid for high temperatures T, by expanding in powers of $1/T$ and keeping only the lowest finite power of $1/T$ which occurs in this expansion. Using this approximation:

(b) Find the magnetizaton M.

(c) If T_i is the temperature of the spin system in a given initial field of magnitude $\mathbf{H}_i$, what is its final temperature T_f if the field is reversibly and adiabatically lowered to the value $\mathbf{H}_f = 0$?

17.4 Calculate the entropy $S = S(\mathcal{H}, T)$ of a paramagnetic diatomic gas according to Langevin's theory and the change of temperature dT accompanied by an infinitesimal change $d\mathcal{H}$ of the magnetic field in an adiabatic process.

17.5 A molecule shall be considered in a state with angular momentum $\hbar\ell$; the projection of its magnetic moment on the z-direction shall be given by

$$\mu_m = mg \qquad ; \quad m = \ell, \ell - 1, \ldots, -\ell$$

Determine the mean value of the total moment $\bar{M}$ of a gas consisting of N molecules as a function of the external field $\mathcal{H}$ in the z-direction and the absolute temperature T, both for integer and half-integer values of ℓ. Show that in the limit in which the product $g\ell = \mu$ is kept finite but where ℓ tends toward infinity, one obtains Langevin's formula for $\bar{M}$.

17.6 Use Hamilton's equations to derive the equations of motion in Eqs.(17.18-17.19) from the hamiltonian in Eq.(17.17).

17.7 Calculate and plot the magnetization at all $\mathcal{H}$ arising from Eq.(17.67), thus generating Figure 17.4.

20.1 Construct an explicit proof of the quantum analog of Liouville's theorem in Eq.(20.30) using an argument similar to that developed for the classical case in Sec. 3.

23.1 The hamiltonian H of a system with one degree of freedom and with coordinate q shall be hermitian and not explicitly dependent upon the time t.

(a) Determine an operator function $O(H)$ of the hamiltonian operator H such that

$$Tr(e^{-\beta H}) = \int dpdq\; e^{-ipq/\hbar} O\; e^{ipq/\hbar}$$

with the integration over both p and q extending from $-\infty$ to $+\infty$.
Hint: Use the relation of completeness for the normalized solutions ϕ_n of the time-independent Schrödinger equation $H\phi_n = E_n\phi_n$.

(b) A kernel $K(q,q',t)$ shall have been found such that

$$\psi(q,t) = \int_{-\infty}^{\infty} K(q,q',t)\psi(q',0)dq'$$

satisfies the time-dependent Schrödinger equation for any wavefunction $\psi(q,0)$ at the time $t = 0$. Given only the function $K(q,q',t)$ and the value of $\tau = \hbar/k_BT$, find the statistical expectation value of q at the temperature T.

25.1 What is the contribution to the specific heat of h hydrogen atoms at room temperature ($T = 293°$) originating from the excitation of electronic levels. (In hydrogen there are n^2 levels with energies $E_n = -E_0/n^2$, where $E_0 = 2.2 \times 10^{-11}$ ergs and where n assumes the values $1, 2, 3, \ldots$)

25.2 In the discussion in this section, and in Secs. 13 and 16, the particles have been labelled with an index $i = 1, 2, \ldots, n$. With *identical*, non-localized systems, one cannot physically distinguish configurations in phase space where only the particle *label* is interchanged (there are $N!$ possibilities here). Thus, in summing over distinct physical configurations in the partition function one should really write

$$\mathcal{Z} = \frac{1}{N!} \int e^{-\beta H}d\lambda \quad ; \text{ identical non-localized systems}$$

The reader can readily verify that this extra factor changes none of the thermodynamic results derived in the text involving derivatives with respect to T or V. It does, for example, change the absolute entropy. Assume, consistent with the discussion in Section 24, that $S_0 \equiv 0$ and derive the result for the ideal gas that

$$S = Nk_B \left\{ lnV - lnN + \frac{3}{2}lnT + \frac{3}{2}ln\left[\frac{2\pi mk_B}{h^2}\right] + \frac{5}{2}\right\}$$

$$= Nk_B \left\{ \frac{5}{2}lnT - lnp + ln\left[\left(\frac{2\pi m}{h^2}\right)^{\frac{3}{2}} k_B^{\frac{5}{2}}\right] + \frac{5}{2}\right\}$$

This is the *Sackur-Tetrode* equation.

Since a crystal at $T \to 0$ has $S = 0$ (Nernst Heat Theorem), one knows the entropy change between this state and the state at very high T and low p where it (and everything) becomes a perfect gas, no matter how complicated the intermediate states and intermediate reversible heat flows.

26.1 Find $\overline{< x^2 >}$ according to quantum statistics for a system with the hamiltonian $H = p^2/2m + m\omega^2 x^2/2$ at the temperature T. Show that for $k_B T >> \hbar\omega$ the result agrees with that obtained from classical statistics.

26.2 In the representation where the hamiltonian $H = p^2/2m + m\omega^2 q^2/2$ of a harmonic oscillator with eigenvalues $(n + 1/2)\hbar\omega$ is diagonal, only the matrix elements

$$\pi_{n,n-1} = \pi^*_{n-1,n} = \sqrt{2mn\hbar\omega}$$

of the operators

$$\pi = p + im\omega q$$
$$\pi^* = p - im\omega q$$

are different from zero. (Check that the commutator $[\pi, \pi^*]$ has the value obtained from $[p, q] = \hbar/i$ and that H, expressed in terms of π and π^*, has the correct matrix form $H_{nn'} = (n + 1/2)\hbar\omega\delta_{nn'}$.) Considering two uncoupled harmonic oscillators with equal mass m and frequencies ω_1 and $\omega_2 = 2\omega_1$, show that the operator

$$C = \pi_1^2 \pi_2^* + \pi_1^{*2} \pi_2$$

commutes with the hamiltonian $H = H_1 + H_2$. With the eigenfunctions g_{20} and g_{01}, pertaining to the degenerate states $n_1 = 2$, $n_2 = 0$ and $n_1 = 0$, $n_2 = 1$ respectively, show further that

$$f_\pm = \frac{1}{\sqrt{2}}(g_{20} \pm g_{01})$$

is an eigenfunction of the operator C, and determine the corresponding eigenvalues $C'_\pm$ of C. (You may try but do not have to generalize these proofs to the case where $\omega_2 = N_2\omega_1/N_1$ for arbitrary integers N_1 and N_2 applied to the degenerate states $n_1 = N_2$, $n_2 = 0$ and $n_1 = 0$, $n_2 = N_1$.)

27a.1 The energy levels due to rotation of a diatomic molecule have the values

$$e_{\ell,m} = \hbar^2 \ell(\ell + 1)/2I$$

where $\ell = 0, 1, 2, \ldots$ and m varies for given ℓ from $-\ell$ to $+\ell$. With T for the temperature and T_{rot}^* a characteristic temperature such that

$$k_B T_{rot}^* = \hbar^2 / 2I$$

show that the contribution to the specific heat per mole of an ideal gas, due to rotation, approaches the classical value in the limit $T/T_{rot}^* \longrightarrow \infty$ and find the lowest-order deviation from this value in powers of T_{rot}^*/T. **Hint**: According to the Euler-MacLaurin formula

$$\sum_{\ell=0}^{\infty} f(\ell) = \int_0^{\infty} f(x)dx + \frac{1}{2}f(0) - \frac{1}{12}f'(0) + \frac{1}{720}f'''(0) + \ldots$$

27a.2 The partition function for the rotational degrees of freedom of A diatomic molecules is given by

$$\mathcal{Z} = \mathcal{Z}_m^A$$
$$\mathcal{Z}_m = \sum_{\ell}(2\ell + 1)e^{-K\ell(\ell+1)}$$

with

$$K = \beta \frac{\hbar^2}{2I}$$

where I is the moment of inertia and ℓ is summed over all allowed states of the molecular system. Calculate in the limits (a) and (b) the contribution of the rotational degrees of freedom to the specific heat c_v for one mole of HD; one mole of para-H_2; one mole of ortho-H_2; and one mole of a thermal equilibrium mixture of ortho- and para-H_2.

(a) Low-temperature limit (K large). Keep only terms $\ell \le 1$ in the partition function.

(b) High-temperature limit. Evaluate c_v to the lowest finite power in $1/T$ occurring in this limit, using the Euler-MacLaurin formula

$$\sum_{m=0}^{\infty} f(m) = \int_0^{\infty} f(x)dx + \frac{1}{2}f(0) - \frac{1}{12}f'(0) + \frac{1}{720}f'''(0) + \ldots$$

27a.3 Derive Eqs.(27.18a) starting from cartesian coordinates.

27a.4 Evaluate Eq.(27.40a) numerically and accurately construct the curve shown in Fig.27.3a.

27b.1 Assuming the relation

$$\nu(k) = \nu^* \left| \sin \frac{\pi k}{2k^*} \right|$$

with $-k^* \leq k \leq +k^*$ between the frequency ν and the magnitude $k = [k_x^2 + k_y^2 + k_z^2]^{1/2}$ of the propagation vector for a soundwave in a solid, the relation between ν^* and k^* is to be chosen such as to lead in the limit $k \longrightarrow 0$ to the values v_L and v_T respectively for the phase velocities of longitudinal and transverse soundwaves. For a temperature $T << \theta_D$ where $\theta_D = h\nu^*/k_B$ and with C_0 proportional to T^3, find the specific heat per unit volume in the form $C = C_0[1 + f(T)]$, keeping only the lowest non-vanishing term of an expansion of $f(T)$ in powers of T. (Numerical constants may be given in the form of definite integrals.)

27b.2 Evaluate Eq.(27.31b) numerically, and accurately construct the curve shown in Fig.27.2b.

27c.1 Determine the entropy s per unit volume of black-body radiation (assuming $s = 0$ for $T = 0$) and the radiation pressue p at the temperature T. Find p in atmospheres if the temperature is sufficiently high to have $k_B T = eV$, where $V = 13.5$ volts is the ionization potential of hydrogen and e is the elementary charge.

27c.2 Given the mean value $E(T)$ of a system as a function of its temperature T, show that in classical as well as in quantum statistics

$$\frac{\Delta E}{E} = \left[-k_B T^2 \frac{d}{dT} \left(\frac{1}{E} \right) \right]^{1/2}$$

where ΔE represents the root-mean-square deviation of the energy from its mean value. Apply this formula to a classical ideal monatomic gas of N molecules and the black-body radiation contained in the volume of 1cm^3. In the latter case, find the approximate temperature for which $\Delta E/E \cong 1$.

27d.1 The energy levels of an atom with a magnetic moment μ in the magnetic field $\mathcal{H}$ have the values

$$e_m = -m\mathcal{H}\mu/j$$

where j is a positive fixed integer and m is an integer varying from $-j$ to $+j$. For n atoms per unit volume, find the magnetic moment M per unit volume of an ideal gas at the temperature T as a function of $\mathcal{H}$ and the susceptibility

$$\chi = \frac{M}{\mathcal{H}}$$

in the limit $\mathcal{H} \longrightarrow 0$. Verify that both results agree in the limit $j \longrightarrow \infty$ with those obtained classically if one assumes each atom to have a magnetic moment of fixed magnitude μ.

27d.2 A system consists of one mole of "atoms" in an external magnetic field. Each atom is characterized by the hamiltonian

$$H = \frac{1}{2m}(\mathbf{p} - \frac{e}{c}\mathbf{A})^2 + \frac{1}{2}m\omega^2\mathbf{r}^2$$

where $\mathbf{A} = (1/2)\mathbf{H} \wedge \mathbf{r} = \mathcal{H}\hat{k}$. The eigenvalues of H are

$$E_{nn'm} = (n + \frac{1}{2})\hbar\omega + (n' + \frac{1}{2})\hbar\omega' - \mu_0\mathcal{H}m$$

where

$$\omega'^2 = \omega^2 + \omega_L^2 \quad ; \quad \omega_L = \frac{e\mathcal{H}}{2mc} \quad ; \quad \mu_0 = \frac{e\hbar}{2mc}$$

and the quantum numbers can have the following values

$$n, n' \quad : \quad 0, 1, 2, \ldots$$
$$m \quad : \quad -n', -n' + 2, \ldots, +n'$$

Evaluate the partition function for the system and calculate:

(a) $M = -\frac{\partial F}{\partial \mathcal{H}}$

(b) $\chi = \frac{M}{\mathcal{H}}\big|_{\mathcal{H}=0}$

(c) χ in the limit $k_B T \ll \hbar\omega$

(d) M in the limit $k_B T \gg \hbar\omega$

(e) M in the strong field approximation

$$\left(\text{i.e.,} \quad \frac{\mathcal{H}\mu_0}{\hbar\omega} \gg 1 \quad ; \quad \frac{\mathcal{H}\mu_0}{k_B T} \gg 1 \quad ; \quad \frac{(\hbar\omega)^2}{k_B T \mathcal{H}\mu_0} \ll 1\right)$$

keeping only the lowest occurring finite power of $0(1/\mathcal{H})$

27d.3 The average expectation value $\overline{<Q>} = Tr(Q\bar{\rho})$ of a time-independent dynamical quantity Q may depend upon the time due to time dependence of $\bar{\rho}$.

(a) Show that

$$\frac{d}{dt}\overline{<Q>} = \frac{1}{\hbar}\overline{<[H, Q]>}$$

(b) Apply this result to the case of a magnetic moment $\boldsymbol{\mu} = \gamma \mathbf{a}$ in a constant magnetic field $\mathbf{H}$ ($\mathbf{a}$ is the angular momentum with the commutation relation $[a_x, a_y] = i\hbar\, a_z$ and its cyclical permutations for the components) to show that the vector $\overline{<\boldsymbol{\mu}>}$ rotates with constant magnitude around the field direction and find the frequency of rotation.

27d.4 Evaluate α in Eq.(27.40d) for a spherical hole in a large uniformly magnetized sample.

27e.1 The state of a system with coordinates q_j ($j = 1, 2, \ldots, n$), all of them varying in a range from the value 0 to the value L, shall be defined by the eigenvalues

$$p_j = 2\pi k_j \hbar / L \qquad (k_j = 0, \pm 1, \pm 2, \ldots)$$

of their conjugate momenta p_j with the normalized eigenfunctions given by

$$\phi_k(q) = \frac{1}{\sqrt{L^n}} \exp\left(\frac{i}{\hbar} \sum_{j=1}^{n} p_j q_j\right)$$

(k stands for the set $k_1, k_2, \ldots, k_n$).

(a) Corresponding to the function $F(p, q)$, consider the operator $F[(\hbar/i)\, \partial/\partial q,\ q]$ and its trace

$$Tr F = \sum_k F_{kk}$$

Going to the limit $L \longrightarrow \infty$, assuming further that $F(p, q)$ can be expanded into a series of powers of p_j and q_j and that it is permissible upon the replacement of p_j by $(\hbar/i)\partial/\partial q_j$ in each term of the series to neglect the commutator $[p_j, q_j]$, show that

$$Tr F = \frac{1}{(2\pi\hbar)^n} \int F(p, q)\, d\lambda$$

with

$$d\lambda = \prod_{j=1}^{n} dp_j dq_j$$

(b) Use the preceding result to establish the connection between the statistical density matrix $\bar{\rho}$ and the classical probability density $\rho(p, q)$ such that the expression

$$\overline{<\phi>} = Tr(\bar{\rho}\phi)$$

for the statistical expectation value of an observable ϕ goes over in the classical limit into the mean value

$$\bar{\phi} = \int \rho(p,q)\phi(p,q) \, d\lambda$$

27e.2 The state of a system of particles with coordinates $q_j (j = 1, 2, \ldots, f)$, where $- L/2 \leq q_j \leq +L/2$ and for which the conjugate momenta have the values $p_j = 2\pi n_j \hbar / L$ with integer n_j, is represented by the normalized eigenfunction

$$\phi_{(nj)}(q_j) = \frac{1}{\sqrt{L^f}} \exp\left(\frac{i}{\hbar} \sum_{j=1}^{f} p_j q_j \right)$$

By going to the limit $L \longrightarrow \infty$, show that the partition function can be written in the form

$$\mathcal{Z}(\beta) = \frac{1}{(2\pi\hbar)^f} \int F(p_j, q_j, \beta) \prod_{j=1}^{f} dp_j dq_j$$

where the integral extends over all variables p_j and q_j from $-\infty$ to $+\infty$ and where

$$F(p_j, q_j, \beta) = \left[\exp\left(\frac{-i}{\hbar} \sum_{j=1}^{f} p_j q_j \right) \right] e^{-\beta H} \left[\exp\left(\frac{i}{\hbar} \sum_{j=1}^{f} p_j q_j \right) \right]$$

with the operator H obtained from the hamiltonian

$$H(p_j, q_j) = \sum_{j=1}^{f} \frac{1}{2m_j} (p_j)^2 + U(q_j)$$

through the replacement of p_j by $(\hbar/i) \, \partial/\partial q_j$. Prove further that in the limit $\hbar \longrightarrow 0$, F reduces to the form

$$F_0 = e^{-\beta H(p_j, q_j)}$$

29.1 Extend the arguments in this section to a two-component system.

29.2 (a) Show that the mean square deviation of the particle number in the grand canonical ensemble is given by

$$\overline{\Delta N^2} = \frac{\partial^2}{\partial \lambda^2} ln \mathcal{Z}_G(\lambda, \beta)$$

(b) Show that

$$\frac{\overline{\Delta N^2}}{N^2} = \frac{k_B T}{V}\kappa$$

where $\kappa = -(1/V)(\partial V/\partial P)_{T,N}$ is the isothermal compressibility.

[**Hint:** Make use of the fact that the free energy is extensive $F = Nf(V/N, T)$].

(c) Show that for a perfect gas

$$\overline{\Delta N^2}/N^2 = 1/N$$

29.3 Establish the result that

$$\Omega(T, V, \mu) = -k_B T \, \ln \mathcal{Z}_G(T, V, \mu)$$

where Ω is the thermodynamic potential

$$\Omega = F - \mu N = E - TS - \mu N = -PV$$

which satisfies the differential relation

$$d\Omega = -S dT - P dV - N d\mu$$

[See, for example, Fetter and Walecka, *Quantum Theory of Many-Particle Systems, op. cit.*, §4.]

30.1 A stationary state t of a particle shall have energy $\varepsilon_t (t = 1, 2, \ldots)$. Assuming that the probability of finding N such identical particles, obeying Fermi statistics, with a total energy E is proportional to $\exp[-(\alpha N + \beta E)]$,

(a) Find the mean square of deviation

$$\overline{\Delta n_t^2} = \overline{(n_t - \bar{n}_t)^2}$$

of the number of particles in the state t in terms of the mean value $\bar{n}_t$.

(b) Find the mean square of deviation

$$\overline{\Delta N^2} = \overline{(N - \bar{N})^2}$$

and the absolute maximum of this expression for a given value $\bar{N}$ of N.

30.2 Determine the entropy per unit volume of a Fermi gas of particles with mass m and density n at a low temperature T. (Besides the arbitrary constant, only the dominant term in the temperature-dependent part shall be kept. Numerical integrals need not be evaluated.)

30.3 Derive Eq.(30.60).

31.1 Consider a two-dimensional Bose gas of particles with mass m at a fixed temperature T and with variable density n (= number of particles per unit area).

(a) Determine the fraction of particles Δ which have a kinetic energy less than E and express the result in terms of the value $\Delta \longrightarrow \delta$ obtained in the limit of very low density. Show that $\Delta \longrightarrow 1$ for $n \longrightarrow \infty$.

(b) Assuming $\delta = e^{-4}$, find the value of n for which $\Delta = 1/2$.

31.2 Consider a two-dimensional Einstein-Bose gas of particles with mass m at the temperature T and with a "density" n, where n = number of particles per unit area.

(a) Find the number per unit area $n(E)$ of particles with a kinetic energy less than E, and compare the result with the corresponding classical expression.

(b) Find the energy E_δ such that a fraction δ of all particles has a kinetic energy above E_δ. (Compare Prob. 31.1.)

(c) For a given fixed value $\delta \ll 1$ and for a given temperature T, find and discuss the limiting behavior for large values of n of the function

$$f(E) = \frac{dn(E)}{dE}$$

in the range of arguments $E \leq E_\delta$.

31.3 Calculate the discontinuity in the slope of the specific heat of an ideal Bose gas at the temperature T_c. Show

$$\Delta \left[\frac{\partial C_V}{\partial T} \right]_{T_c} \cong -3.66 \frac{N k_B}{T_c}$$

31.4 Show that at high T, non-interacting Bose and Fermi gases both reproduce the result in Prob. 25.2. (Discuss the role of spin degeneracy.)

A.1 Consider the motion of a point in phase space

$$x_\mu = x_\mu(x_0, t)$$

where the dynamics is governed by Hamilton's Eqs.(A.16) with a hamiltonian $H(x)$. Here the *initial values* are given by

$$x_\mu = \delta_{\mu\kappa}(x_0)_\kappa \qquad ; \text{ for } t = t_0$$

Now define the transformation

$$X_\mu \equiv x_\mu$$

which depends parametrically on the time.

(a) Derive the following first-order differential equation in the time for the Poisson bracket $[X_\mu, X_\nu]_{(x_0)}$

$$\frac{d}{dt}[X_\mu, X_\nu]_{(x_0)} = \frac{\partial}{\partial t}[X_\mu, X_\nu]_{(x_0)}$$

$$= \frac{\partial^2 H}{\partial x_\rho \partial x_\sigma} \left([X_\sigma, X_\nu]_{(x_0)}\varepsilon_{\rho\mu} + [X_\mu, X_\rho]_{(x_0)}\varepsilon_{\sigma\nu}\right)$$

Here the Poisson brackets and partial time derivative are evaluated by keeping the other variables in the set (x_0, t) fixed.

(b) Derive the initial condition

$$[X_\mu, X_\nu]_{(x_0)} = -\varepsilon_{\mu\nu} \qquad ; \text{ for } t = t_0$$

(c) Show that the following expression

$$[X_\mu, X_\nu]_{(x_0)} = -\varepsilon_{\mu\nu}$$

is the solution to these equations for all subsequent times.

(d) Hence, conclude that the *time development* of this dynamical system represents a *canonical* transformation.

B.1 Verify the identity in Eq.(B.19) explicitly for a 2×2 and a 4×4 matrix.

Index

Index

Italic numbers denote illustrations.

Library of Congress Cataloging-in-Publication Data

Bloch, Felix, 1905–
Fundamentals of statistical mechanics : manuscript and notes of Felix Bloch / prepared
by John Dirk Walecka.
    p.   cm.
Includes index.
ISBN 0-8047-1501-7 (alk. paper) :
    1. Statistical mechanics.   I. Walecka, John Dirk, 1932–
II. Title.
QC174.8.B59   1989   88-12285
530.1'3—dc19   CIP